State Estimation and Fault Diagnosis under Imperfect Measurements

The objective of this book is to present the up-to-date research developments and novel methodologies on state estimation and fault diagnosis (FD) techniques for a class of complex systems subject to closed-loop control, nonlinearities, and stochastic phenomena. It covers state estimation design methodologies and FD unit design methodologies including the framework of optimal filter and FD unit design, robust filter and FD unit design, stability, and performance analysis for the considered systems subject to various kinds of complex factors.

Features:

- Reviews latest research results on the state estimation and fault diagnosis issues.
- Presents comprehensive framework constituted for systems under imperfect measurements.
- Includes quantitative performance analyses to solve problems in practical situations.
- Provides simulation examples extracted from practical engineering scenarios.
- Discusses proper and novel techniques such as the Carleman approximation and completing the square method is employed to solve the mathematical problems.

This book aims at Graduate students, Professionals and Researchers in Control Science and Application, Stochastic Process, Fault Diagnosis, and Instrumentation and Measurement.

State Estimation and Fault Diagnosis under Imperfect Measurements

Yang Liu
Zidong Wang
Donghua Zhou

CRC Press
Taylor & Francis Group
Boca Raton London New York

CRC Press is an imprint of the
Taylor & Francis Group, an **informa** business

First edition published 2023
by CRC Press
6000 Broken Sound Parkway NW, Suite 300, Boca Raton, FL 33487-2742

and by CRC Press
4 Park Square, Milton Park, Abingdon, Oxon, OX14 4RN

CRC Press is an imprint of Taylor & Francis Group, LLC

© 2023 Yang Liu, Zidong Wang and Donghua Zhou

Reasonable efforts have been made to publish reliable data and information, but the author and publisher cannot assume responsibility for the validity of all materials or the consequences of their use. The authors and publishers have attempted to trace the copyright holders of all material reproduced in this publication and apologize to copyright holders if permission to publish in this form has not been obtained. If any copyright material has not been acknowledged please write and let us know so we may rectify in any future reprint.

Except as permitted under U.S. Copyright Law, no part of this book may be reprinted, reproduced, transmitted, or utilized in any form by any electronic, mechanical, or other means, now known or hereafter invented, including photocopying, microfilming, and recording, or in any information storage or retrieval system, without written permission from the publishers.

For permission to photocopy or use material electronically from this work, access www.copyright.com or contact the Copyright Clearance Center, Inc. (CCC), 222 Rosewood Drive, Danvers, MA 01923, 978-750-8400. For works that are not available on CCC please contact mpkbookspermissions@tandf.co.uk

Trademark notice: Product or corporate names may be trademarks or registered trademarks and are used only for identification and explanation without intent to infringe.

ISBN: 978-1-032-31385-6 (hbk)
ISBN: 978-1-032-31386-3 (pbk)
ISBN: 978-1-003-30948-2 (ebk)

DOI: 10.1201/9781003309482

Typeset in CMR10
by KnowledgeWorks Global Ltd.

To our families and our friends.

Contents

List of Figures	xi
List of Tables	xiii
Preface	xv
Author's Biography	xvii
Acknowledgment	xxi
Symbols	xxiii

1 Introduction   1
- 1.1 Challenges with Imperfect Measurements   1
  - 1.1.1 Measurement Noises   1
  - 1.1.2 Nonlinear Measurements   1
  - 1.1.3 Communication Constraints   2
- 1.2 Analysis and Synthesis of Imperfect Measurements   4
  - 1.2.1 Measurement Noises   4
  - 1.2.2 Nonlinear Measurement   6
  - 1.2.3 Communication Constraints   7
    - 1.2.3.1 Transmission Delay   7
    - 1.2.3.2 Missing Measurement   9
    - 1.2.3.3 Signal Quantization 11
    - 1.2.3.4 Sensor Gain Degradation 14
    - 1.2.3.5 Event-Triggered Transmission 15
    - 1.2.3.6 Sensor Saturation 19
    - 1.2.3.7 Integral Measurement 22
- 1.3 Outline of This Book 22

2 Optimal Filtering for Networked Systems with Stochastic Sensor Gain Degradation   27
- 2.1 Problem Formulation and Preliminaries 28
- 2.2 Optimal Filter Design 30
- 2.3 Simulation Example 36
- 2.4 Conclusions 37

3 Recursive Filtering over Sensor Networks with Stochastic Sensor Gain Degradation — 39
- 3.1 Problem Formulation and Preliminaries — 41
- 3.2 Main Results — 43
  - 3.2.1 Filter Design — 43
  - 3.2.2 Boundedness — 46
  - 3.2.3 Monotonicity — 51
- 3.3 Numerical Example — 56
- 3.4 Conclusions — 59

4 H_∞ Filtering for Nonlinear Systems with Stochastic Sensor Saturations and Markov Time Delays — 61
- 4.1 Problem Formulation — 62
- 4.2 Main Results — 65
- 4.3 Simulation Examples — 74
  - 4.3.1 Example A — 74
  - 4.3.2 Example B — 75
- 4.4 Conclusion — 78

5 Observer Design for Systems with Unknown Inputs and Missing Measurements — 79
- 5.1 Problem Formulation — 80
- 5.2 Observer Design — 81
- 5.3 Boundedness Analysis — 82
- 5.4 Illustrative Examples — 85
- 5.5 Conclusions — 87

6 Filtering and Fault Detection for Nonlinear Systems with Polynomial Approximation — 89
- 6.1 Problem Formulation — 91
  - 6.1.1 Polynomial Approximation of Nonlinear Functions — 92
  - 6.1.2 The Polynomial Nonlinear Systems — 94
  - 6.1.3 The Filter and the Fault Detection Problems — 96
- 6.2 Polynomial Filter Design — 97
- 6.3 Fault Detection — 103
- 6.4 Illustrative Example — 108
- 6.5 Conclusion — 113

7 Event-Triggered Filtering and Fault Estimation for Nonlinear Systems with Stochastic Sensor Saturations — 115
- 7.1 Problem Formulation — 116
- 7.2 Filter Design — 118
- 7.3 Boundedness Analysis — 123
- 7.4 Fault Estimation — 127
- 7.5 Illustrations — 129
- 7.6 Conclusions — 132

8 Finite-Horizon Quantized H_∞ Filter Design for Time-Varying Systems under Event-Triggered Transmissions — 133
- 8.1 Problem Formulation 134
- 8.2 Filter Design 137
- 8.3 An Illustrative Example 144
- 8.4 Conclusion 148

9 Observer-Based Fault Diagnosis Schemes under Closed-Loop Control — 149
- 9.1 Unknown-Input-Observer Method 150
- 9.2 Luenberger-Observer-Based and Robust-Observer-Based Methods 155
- 9.3 A Simulation Example 159
- 9.4 Conclusion 162

10 State Estimation and Fault Reconstruction with Integral Measurements under Partially Decoupled Disturbances — 163
- 10.1 Problem Formulation 165
- 10.2 Filter Design 167
- 10.3 Parameter Calculation 169
- 10.4 Illustrative Example 171
- 10.5 Conclusion 175

11 Conclusion and Further Work — 177

Bibliography — 179

Index — 197

List of Figures

2.1	Trace of estimation error variances for the states	37
3.1	The state x_1 and its estimate	57
3.2	The state x_2 and its estimate	58
3.3	The estimation error and bound	58
3.4	The estimation error and sensor gain degradation	59
4.1	Nonlinear filtering performance	75
4.2	Delay-dependent and delay-independent filtering performances	77
4.3	Delay-dependent and delay-independent filtering errors	78
5.1	The values of α_k	85
5.2	The state x_1 and its estimate	86
5.3	The state x_2 and its estimate	86
6.1	The state x_1 and its estimate	112
6.2	The state x_2 and its estimate	112
6.3	The Euclidean norm of residual and the threshold	113
7.1	The state x_1 and its estimate	130
7.2	The state x_2 and its estimate	130
7.3	The estimation error and bound	131
7.4	The fault and its estimate	131
8.1	The state x_1 and its estimate	146
8.2	The state x_2 and its estimate	146
8.3	The state x_3 and its estimate	147
8.4	The Euclidean norm of the residual	147
9.1	Residual obtained with (9.13)	161
9.2	Residual obtained with (9.22)	161
10.1	The state x_1 and its estimate	173
10.2	The state x_2 and its estimate	173
10.3	The state x_3 and its estimate	174
10.4	The actual additive fault and its estimate	174

List of Tables

5.1 Estimation Performance Comparison 87

6.1 Average Trace of the Estimation Error Covariance 111

Preface

With the increasing scale and complexity of modern industrial systems, how to improve the reliability and safety of practical systems and eliminate potential threats has been extensively studied. As a result, it is of great significance to monitor the systems effectively. State estimation and fault diagnosis (FD) techniques enable us to better understand the working condition of addressed systems and prevent severe performance deteriorations as early as possible. In practical engineering, many systems are subject to imperfect measurement such as disturbance, nonlinearity, stochasticity, and so on. These phenomena have a great impact on the performance of the state estimation and FD, and the traditional estimation/filtering and FD schemes may be no longer applicable in these cases. Consequently, it is important to redesign the filter/FD unit or even develop new methodologies under imperfect measurements. The disturbance can influence the accuracy of state estimation and FD results. The nonlinearities, if not adequately dealt with, may lead to undesirable dynamic behaviors such as oscillation or even instability. The stochastic characteristics are especially obvious in networked systems connected via wireless and shared links. These factors pose extra challenges to state estimation and FD problems.

The objective of this book is to present the up-to-date research developments and novel methodologies on state estimation and FD for a class of systems subject to disturbances, nonlinearities, and stochastic phenomena. The content of this book can be divided into two parts, where the first part (Chapters 2–5) presents state estimation design methodologies and the second part (Chapters 6–10) shows the FD unit design methodologies. The work provides a framework of optimal filter/FD unit design, robust filter/FD unit design, stability and performance analysis for the considered systems subject to various kinds of complex factors in measurements, including missing measurements, quantizations, event-triggered transmissions, etc. Several techniques including recursive Riccati equations, matrix decomposition, optimal estimation theory, and mathematical optimization methods are employed to develop the desired filter/FD unit. In addition, this book provides valuable reference materials for researchers who wish to explore the state estimation and FD in these complicated cases.

The concise frame and description of the book are given as follows. Chapter 1 introduces the recent advances on state estimation/filtering and FD problems under imperfect measurements and the outline of the book. Chapter 2 is concerned with the optimal filtering for networked systems with stochastic sensor gain degradation. Chapter 3 studies the minimum-variance recursive filtering over sensor networks with stochastic sensor gain degradation, where the filtering performance is analyzed with respect to the boundedness and monotonicity. Chapter 4 considers the H_∞ filtering for nonlinear systems with stochastic sensor saturations and Markov time delays. Chapter 5 deals with observer design problem for systems with unknown inputs and missing measurements, and establishes the sufficient conditions for the uniform ultimate boundedness of the filtering error systems. Chapter 6 addresses the filtering and fault detection problems for nonlinear systems with polynomial approximation. Chapter 7 copes with the event-triggered filtering and fault estimation problems for nonlinear systems with stochastic sensor saturations. In Chapter 8, a finite-horizon quantized H_∞ filter is designed for a class of time-varying systems under event-triggered transmissions. Chapter 9 discusses a class of observer-based fault diagnosis schemes under closed-loop control. Chapter 10 is concerned with the state estimation and fault reconstruction with integral measurements under partially decoupled disturbances. Chapter 11 gives the conclusion and some possible future research directions. Simulations presented in this book are implemented using The MathWorks MATLAB software package.

This book is a research monograph whose intended audience is graduate and postgraduate students as well as researchers. The background required of the reader is knowledge of basic stochastic process, basic control system theory, basic Lyapunov stability theory, and basic optimal estimation theory.

Yang Liu
Loughborough, U.K.

Zidong Wang
London, U.K.

Donghua Zhou
Qingdao, China

Author's Biography

Yang Liu received the B.Sc. degree and the Ph.D. degree in the Department of Automation at Tsinghua University, Beijing, China, in 2010 and 2016, respectively. From August 2016 to September 2018, he worked as a Postdoctoral Researcher in the College of Electrical Engineering and Automation, Shandong University of Science and Technology, Qingdao, China. Then he worked in the same department as a Lecturer and was promoted to an Associate Professor in January 2020. In April 2021, he joined the Department of Aeronautical and Automotive Engineering, Loughborough University, U.K., as a Research Associate. He has published over 40 papers in refereed international journals. His research interests include optimal filtering, complex systems, neural networks as well as fault detection and diagnosis. He is a very active reviewer for several international journals. He was an outstanding reviewer for IEEE Transactions on Automatic Control in 2013, for the journal Neurocomputing in 2016, and for Journal of the Franklin Institute in 2016. He was a guest editor for Systems Science and Control Engineering in 2020.

Zidong Wang is a Professor of Dynamical Systems and Computing at Brunel University London, West London, United Kingdom. He was born in 1966 in Yangzhou, Jiangsu, China. He received the B.Sc. degree in Mathematics in 1986 from Suzhou University, Suzhou, the M.Sc. degree in Applied Mathematics in 1990 and the Ph.D. degree in Electrical and Computer Engineering in 1994, both from Nanjing University of Science and Technology, Nanjing.

He was appointed as Lecturer in 1990 and Associate Professor in 1994 at Nanjing University of Science and Technology. From January 1997 to December 1998, he was an Alexander von Humboldt research fellow with the Control Engineering Laboratory, Ruhr-University Bochum, Germany. From January 1999 to February 2001, he was a Lecturer with the Department of Mathematics, University of Kaiserslautern, Germany. From March 2001 to July 2002, he was a University Senior Research Fellow with the School of Mathematical and Information Sciences, Coventry University, U.K. In August 2002, he joined the Department of Computer Science, Brunel University London, U.K., as a Lecturer, and was then promoted to a Reader in September 2003 and to a Chair Professor in July 2007.

Professor Wang's research interests include dynamical systems, signal processing, bioinformatics, control theory and applications. He has published more than 600 papers in refereed international journals. He was awarded the Humboldt research fellowship in 1996 from Alexander von Humboldt

Foundation, the JSPS Research Fellowship in 1998 from Japan Society for the Promotion of Science, and the William Mong Visiting Research Fellowship in 2002 from the University of Hong Kong. He was a recipient of the State Natural Science Award from the State Council of China in 2014 and the Outstanding Science and Technology Development Awards (once in 2005 and twice in 1997) from the National Education Committee of China.

Professor Wang is currently serving or has served as the Editor-in-Chief for International Journal of Systems Science, the Editor-in-Chief for Neurocomputing, Executive Editor for Systems Science and Control Engineering, Subject Editor for Journal of The Franklin Institute, an Associate Editor for IEEE Transactions on Automatic Control, IEEE Transactions on Control Systems Technology, IEEE Transactions on Systems, Man, and Cybernetics - Systems, Asian Journal of Control, Science China Information Sciences, IEEE/CAA Journal of Automatica Sinica, Control Theory and Technology, an Action Editor for Neural Networks, an Editorial Board Member for Information Fusion, IET Control Theory & Applications, Complexity, International Journal of Systems Science, Neurocomputing, International Journal of General Systems, Studies in Autonomic, Data-driven and Industrial Computing, and a member of the Conference Editorial Board for the IEEE Control Systems Society. He served as an Associate Editor for IEEE Transactions on Neural Networks, IEEE Transactions on Systems, Man, and Cybernetics - Part C, IEEE Transactions on Signal Processing, Circuits, Systems & Signal Processing, and an Editorial Board Member for International Journal of Computer Mathematics.

Professor Wang is a Member of the Academia Europaea (section of Physics and Engineering Sciences), a Fellow of the IEEE (for contributions to networked control and complex networks), a Fellow of the Chinese Association of Automation, a Member of the IEEE Press Editorial Board, a Member of the EPSRC Peer Review College of the UK, a Fellow of the Royal Statistical Society, a member of program committee for many international conferences, and a very active reviewer for many international journals. He was nominated an appreciated reviewer for IEEE Transactions on Signal Processing in 2006-2008 and 2011, an appreciated reviewer for IEEE Transactions on Intelligent Transportation Systems in 2008, an outstanding reviewer for IEEE Transactions on Automatic Control in 2004 and for the journal Automatica in 2000.

Donghua Zhou received the B.Eng., M.Sci., and Ph.D. degrees in electrical engineering from Shanghai Jiaotong University, Shanghai, China, in 1985, 1988, and 1990, respectively.

He was an Alexander von Humboldt Research Fellow with the University of Duisburg, Duisburg, Germany, from 1995 to 1996, and a Visiting Scholar with Yale University, New Haven, CT, USA, from 2001 to 2002. He joined Tsinghua University, Beijing, China, in 1996, and was promoted as a Full Professor in 1997, he was the Head of the Department of Automation, Tsinghua University, during 2008 and 2015. He is now a Vice President, Shandong University of Science and Technology, Qingdao, China, and a Joint Professor with

Tsinghua University. He has authored and coauthored over 230 peer-reviewed international journal papers and 7 monographs in the areas of fault diagnosis, fault-tolerant control, and operational safety evaluation.

Dr. Zhou is a Fellow of CAA and IET, a Member of IFAC TC on SAFEPROCESS, an Associate Editor of Journal of Process Control, the Vice Chairman of Chinese Association of Automation (CAA), the TC Chair of the SAFEPROCESS committee, CAA. He was also the NOC Chair of the 6th IFAC Symposium on SAFEPROCESS 2006.

Acknowledgment

This book would not have been possible without the help, support and guidance of many people. The authors would like to express their deep appreciation to those who have been directly involved in various aspects of the research leading to this book.

Special thanks go to Professor Xiao He from Tsinghua University for his valuable suggestions, constructive comments, and support. The authors also extend our thanks to many colleagues who have offered support and encouragement throughout this research effort. In particular, we would like to acknowledge the contributions from Xiaosheng Si, Muheng Wei, Rongyi Yan, Qinyuan Liu, Jianxun Zhang, Tianxu Guo, Man Xu, Lei Zou, Hongli Dong, Derui Ding, Jun Hu, Lihui Cui, Zhongyi Zhao, and Jiyue Guo. Last but not the least, the authors are especially grateful to their families for their encouragement and never-ending support when it was most required.

The writing of this book was supported by National Natural Science Foundation of China (NSFC) under Grants 61703244, 61733009, 61933007, 61873148, 61751307, 61773400, and Research Fund for the Taishan Scholar Project of Shandong Province of China.

Symbols

$\odot$	The Hadamard product
$\otimes$	The Kronecker product
$\mathbb{R}^n$	The n-dimensional Euclidean space.
$\mathbb{R}^{n \times m}$	The set of all $n \times m$ real matrices.
$\mathbb{R}^+$	The set of all positive real numbers
$\mathbb{N}$	The set of natural numbers.
$\mathbb{S}_+^n$	The set of $n \times m$ positive definite matrices.
A^T or A'	The transpose of matrix A.
$A^\dagger$	The Moore-Penrose pseudo inverse of A
$A > 0$	The matrix A is positive definite.
$A \geq 0$	The matrix A is positive semidefinite.
$A < 0$	The matrix A is negative definite.
$A \leq 0$	The matrix A is negative semidefinite.
$\|\cdot\|$	The Euclidian norm of real vectors or the spectral norm of real matrices.
$\|\cdot\|_{\min}$	The smallest singular value of a matrix.
$\mathrm{tr}(A)$	The trace of matrix A.
$\|x\|_P^2$	Equals to $x^T P x$ when x is a vector.
$\mathbb{P}\{\cdot\}$	The occurrence probability of the event "$\cdot$".
$\mathbb{E}\{x\}$	The expectation of stochastic variable x.
$\mathrm{Var}\{x\}$	The variance of stochastic variable x.
$\mathbb{E}\{x\|y\}$	The conditional expectation of x given y.
I	The identity matrix of compatible dimension.

0	The zero matrix of compatible dimension.
$\mathbf{0}_n$	The $n \times n$ zero matrix.
$\mathbf{1}_n$	The $n \times 1$ column vector with all elements equal to 1.
$\text{vec}\{x_1, x_2\}$	The column vector $\begin{bmatrix} x_1^T & x_2^T \end{bmatrix}^T$.
$\text{vec}_n\{x_i\}$	The column vector $\text{vec}\{x_1^T, x_2^T, \cdots x_n^T\}$.
$\text{diag}\{x_1, x_2\}$	The block diagonal matrix with ith block being x_i and all other entries being zero.
$\text{diag}_n\{A_i\}$	The block diagonal matrix $\text{diag}\{A_1, A_2, \cdots, A_n\}$.
$\{M_{ij}\}_{n \times n}$	The partitioned matrix with M_{ij} being (i, j)-th block submatrix.
$\mathcal{L}_2([0, T]; \mathbb{R}^n)$	The space of square-summable n-dimensional vector functions over $[0, T]$.

1
Introduction

1.1 Challenges with Imperfect Measurements

1.1.1 Measurement Noises

In classical automation disciplines, the perfect measurements are linear combination of the system states. Unfortunately, in practical engineering, sensors are mostly contaminated by the ubiquitous noises introduced by environmental disturbances, variations of sensor parameters, unmodeled system dynamics, and some other factors. For the state estimation issue, how to attenuate the effects of noises is a mainstream theme and for the FD topic, how to make the residual signal sensitive to faults and robust to noises and uncertainties simultaneously is a major task. In stochastic systems, measurement noises have mostly been assumed to be white Gaussian noises, and the optimal state estimator and FD unit can be established by resorting to the mean and variance of the noises in the minimum variance sense. When the noises are coloured or the distribution is non-Gaussian, the cross-correlations between noises in different time steps and some stochastic characteristics of distributions have been used to parameterize the filter and FD unit. In the deterministic framework, square-summable noises have been handled in the H_∞ sense, and the linear matrix inequality (LMI) technique and its improvement have been frequently utilized. For the amplitude-bounded noises, set-membership-based estimation and diagnosis have been widely employed.

1.1.2 Nonlinear Measurements

Almost all the practical systems are nonlinear, so it is of great significance to consider nonlinear measurements in the state estimation and FD studies. However, the nonlinear measurements can deteriorate the system observability and pose extra challenges to the analysis and synthesis of the system. In many cases, the nonlinear functions in the measurement equation have been linearized such that the mature linear methods can be applied. By neglecting the high-order linearization error or introducing some linear constraints such as Lipschitz condition and sector-bounded condition, the nonlinear measurement can be dealt with in a linear framework as well. Naturally, this kind of strategy will unavoidably lead to some conservativeness. When the

DOI: 10.1201/9781003309482-1

nonlinearities are directly considered, one of the key difficulties is the possibility of nonlinear functions going to infinity in a finite time. This is remarkably intractable if the convergency and stability of the proposed estimation and FD algorithms need to be analyzed.

1.1.3 Communication Constraints

In traditional control theory, the bandwidth of the transmission link is assumed to be infinite, which means that the received signal mostly just contains the valid information and some additive noises, and the accuracy of the transmitted signal can be arbitrarily high. This assumption, unfortunately, does not always hold in reality scenarios. For example, in a formation consisting of several unmanned aerial vehicles, the information between them can only be exchanged through unsteady wireless links with limited bandwidth. As a result, the signals transmitted in these cases may be subject to many undesired phenomena such as disorder, time delay, missing measurement, etc. These situations naturally pose extra challenges to the traditional control theory. The classical methods must be modified or redesigned to cope with these imperfect measurements, and sometimes they may even be not applicable any longer. Therefore, much research attention has been devoted to handling the measurements transmitted recently, especially scholars focusing on networked environments. Some specific phenomena brought in by communication constraints are given as follows.

- **Transmission delay.** Since a practical signal cannot be received as soon as it is transmitted and signal processing takes some time in every component, time delay is frequently encountered in real systems, especially when the system is spatially distributed or the processing speed of the devices is slow, such as chemical and biological processes, hydraulic systems, and manufacturing processes. Both the sensor-to-controller and the controller-to-actuator delay can limit and degrade the performance of addressed systems and even induce instability. Delays can be classified into discrete delays, distributed delays, neutral delays, and so on.

- **Missing measurement.** Missing measurement, which is also referred to as package dropout, mainly results from the transmission congestion in the links. When the data transmitted through the device or link exceeds its capacity, congestion follows and the receiving end may fail to obtain the transmitted signal.

- **Signal quantization.** In reality, the accuracy of the transmitted data cannot be infinitely high. As a result, when analog signals need to be transmitted, a quantizer is usually deployed with hope to realize the analog/digital conversion. The quantized signals are with a finite word length, which naturally brings in some information loss.

- **Sensor gain degradation.** Sensor gain variation occurs frequently in engineering practice. This is particularly true for real-world systems under changeable working conditions. It is quite common in a networked system that the sensor gains degrade in a random fashion, which is due probably to sensor aging, intermittent sensor outages, or network-induced saturations/congestions.

- **Event-triggered transmission.** In the past few decades, the event-triggered transmission mechanism arouses a great deal of interest due to the rapid development of computer science and digital microprocessor. Compared with the conventional clock-driven strategy referring to periodic signal transmissions, in an event-triggered scheme, the outputs/inputs are released only when some conditions are violated. By reducing signal exchanges, the event-triggered could avoid some harmful transmission phenomena (e.g. data dropout, time delay, and congestion), improve the energy efficiency and extend the lifetime of the services. But at the same time, the event-triggered transmission will naturally lead to intermittent transmission.

- **Sensor saturation.** Due to physical and technological limitations, sensors cannot provide signals with unbounded amplitudes and such saturation phenomena pose extra challenges to the systems design. Sensors in practical systems might frequently encounter some transient phenomena especially when systems are deployed in unsteady environments. Under the circumstances, the saturation itself may undergo random switches/changes in its occurrence/intensity because of various reasons such as random sensor failures and abrupt environmental changes.

- **Integral Measurement.** In reality, the system measurements might be in proportion to the integral of the system states over a given time period due to the delayed data collection and the real-time signal processing. This phenomenon, known as integral measurement, often occurs in engineering applications such as chemical processes and nuclear reaction processes.

It is worth mentioning that there are some other communication constraints, especially in networked environments, such as fading channels, transmission protocols (try-once-discard protocol, Round-Robin protocol, stochastic protocol, etc), coding-decoding schemes, and so on. The constraints emphasized above are the main phenomena which will be discussed in the book.

1.2 Analysis and Synthesis of Imperfect Measurements

1.2.1 Measurement Noises

In a disturbance-free situation, filters and observers have been designed for linear systems to cope with the error induced by the inaccurate initial conditions, such as the classical Luenberger observer proposed in 1971 [109]. In these cases, the state estimation error can tend to be zero when time approaches infinity, so can the residual signal, which is widely employed in the FD study and often taken as the difference (or its linear combination) between the actual and the predicted measurement. Unfortunately, disturbances are ubiquitous in practice, and the estimation/FD results are almost always subject to the effects of disturbances. For disturbances with different properties, the following strategies have been proposed correspondingly.

- **Stochastic disturbances.** To cope with stochastic disturbances, classical Kalman filter has proven to be an effective way to establish the optimal filter in the minimum variance sense for linear systems with Gaussian white noises [77]. With the covariances of the Gaussian noises, the estimation error covariance has been calculated and minimized via selecting a proper filter gain at each time step.

    Improved Kalman filter has been extensively studied in the last decades. Extended Kalman filter (EKF) has been used to deal with nonlinear systems [27], where the state transition matrix and the measurement matrix have been replaced by the linearization coefficients and the linearization errors have been neglected. The unscented Kalman filter (UKF) can reduce the linearization errors of the EKF [75]. In UKF, the statistical information of disturbances has been utilized and the unscented transformation has been employed to approximate the nonlinear dynamics. Experience has shown that the numerical solution of the estimation error covariance in Kalman filter often leads to a covariance matrix with a negative characteristic root, which can stop the normal execution of the algorithm. This condition can be avoided by calculating a "square root" of the matrix, and the technique has often been referred to as square root Kalman filter [7]. When the addressed system is large-scale or spatially distributed, a kind of distributed Kalman filter has been developed [80], where the observations have been fused based on the estimation results of the local Kalman filters using bipartite fusion graphs and consensus averaging algorithms. There have been some other improved Kalman filters such as adaptive Kalman filter (AKF) which can deal with unavailable disturbance statistics [10].

    It is noted that the estimation performance of the Kalman filter and its improvements has been a hot research topic for a long time as well. For a linear system, the boundedness of state estimation error covariance with Kalman filter can be guaranteed by some complete uniform observability and complete uniform controllability conditions related to the state transition matrix, measurement matrix, and the covariance of the additive plant noise [29]. The convergence and stability of EKF have been investigated with special forms of the state equations [42,149]. In 1999, Reif et al. have studied the exponential boundedness in mean square and the boundedness with probability one of the estimation error covariance by resorting to the Lyapunov theory and presented several sufficient conditions which can guarantee the boundedness [136].

- **Square-summable disturbances.** For square-summable disturbances, the H_∞ filter has been widely used which can provide a bound for the worst-case estimation error without the knowledge of noise statistics. There have mostly been two techniques employed to characterize the H_∞ filter: LMI technique [43,60,173] and the Riccati differential/difference equation (RDE) technique [4,5,50]. Originally, LMI is only applicable to time-invariant systems where a static filter can be designed. To deal with the time-varying systems, the so-called differential/difference linear matrix inequality (DLMI) and recursive linear matrix inequality (RLMI) methods [31,140,143] have been developed to effectively solve the finite-horizon H_∞ filtering problems.

When it comes to the FD problem, the H_∞ criterion can be applied based on the following two strategies. The bound of the residual signal or the fault estimation error with respect to the disturbance has been investigated in the H_∞ framework when the fault needs to be directly estimated or the residual needs to indicate the occurrence of the possible fault [3,111,205]. An alternative approach is to design the residual signals by resorting to the H_-/H_∞ criterion, which can minimize the effects of disturbances while maximize the effects of faults [96,141,199,203].

- **Norm-bounded disturbances.** For norm-bounded disturbances (also known as unknown-but-bounded disturbances), there have been mainly two kinds of estimation approaches: the interval observer and the set-membership estimation. The interval observer approach aims to obtain upper and lower bounds of estimated states using two observers. The original interval observer has been designed in [113], and an interval observer in the form of two Luenberger observers has been constructed, the structure of which has been widely adopted in the following results. The interval observer has been designed in [38,134], where the estimation error has also been analyzed. Furthermore, the performance optimization problem of the interval observer in the L_1/L_2 sense has been discussed in [21]. The set-membership estimation scheme has been another widely adopted method to cope with the time-varying systems with norm-bounded noises. The goal of the set-membership estimation is to calculate a compact set containing the true state of the system based on the system model and the bounds of the disturbances. Since proposed in [139,170], set-membership estimation has been widely used in the state estimation [103,119,197] due to its merits of eliminating wrapping effects via an iterative approach and low computation load in efficiently calculating the Minkowski sum.

Both the interval observer and the set-membership estimation are applicable to the fault detection problem since the calculated bounds can be used to determine the threshold for the residual signal. The interval observer has been applied to the fault detection problems for relatively complex systems such as switched systems [152, 188] and multi-agent systems [196]. The set-membership-based fault diagnosis technique has also been adopted in many systems such as DC motor [165], electrical connectors [70], and mobile robots [201]. In fact, these two approaches have been compared in the fault detection issue for uncertain systems [127].

It is noted there are some other state estimation and FD methods that can cope with different kinds of disturbances. For example, unknown input observer (UIO) aims to decouple the disturbances so that the effects of the disturbances on the estimation/FD results can be completely eliminated. Since it was proposed in 1996 [23], the UIO technique has been widely adopted as a renowned disturbance decomposition approach. Under certain assumptions on the measurement matrix and the disturbance distribution matrix, the desired decoupling relationship can be realized in the UIO.

1.2.2 Nonlinear Measurement

The nonlinear issue has been attracting researchers from a variety of disciplines. As discussed in the previous subsection, the famous EKF algorithm has proved to effective to solve the estimation problem for nonlinear systems in the least mean square sense. Recently, considerable attention has been paid to the performance improvement of the traditional EKF with respect to the insensitivity to the parameter uncertainties as well as the capability of handling nonlinearities [73, 179]. Polynomial extended Kalman filter (PEKF) is an extension of EKF with aim to cater for inherent nonlinearies using polynomial approximations. Traditional EKF is only concerned with the *linear* term and simply ignores the linearization error, while PEKF considers the Carleman approximation of a nonlinear system of a given order μ [85]. The order could be determined according to the form of the nonlinearity and the estimation performance specification. In this sense, the PEKF is more applicable than EKF as far as the accuracy is concerned. When the order $\mu = 1$, PEKF reduces to conventional EKF. A PEKF is designed to cope with an augmented state which is made of Kronecker powers of the original state [52]. Due to its higher accuracy than that of EKF, the PEKF has stirred quite a lot research attention and many corresponding results have been reported in the literature [51].

In many cases, the addressed nonlinear functions have mostly been assumed to satisfy some linear constraints. For example, the Lipschitz condition is often formulated as follows:

The nonlinear function f is said to satisfy Lipschitz condition if there exists a known real matrix M such that

$$\|f(x) - f(y)\| \leq \|M(x - y)\|. \tag{1.1}$$

Another widely adopted linear constraint for nonlinear function is the sector-bounded condition:

The nonlinear function f is said to satisfy sector-bounded condition if there exist known real matrices M_1 and M_2 such that

$$[f(x) - f(y) - M_1(x - y)]^T [f(x) - f(y) - M_2(x - y)] \leq 0. \tag{1.2}$$

It can be seen that under these conditions, the nonlinear function can be handled in a linear framework, and much research attention has been paid to deal with nonlinear systems under Lipschitz condition [120, 175, 184] and sector-bounded condition [1, 24, 110]. It is noted that there are some other ways to constrain the addressed nonlinear functions with linear conditions such as the ellipsoid-bounded condition [171].

The above mentioned results have been obtained when the addressed nonlinear function is known to us. In reality, it is also often the case that the nonlinearity is complex or even impossible to accurately model. In such a situation, many functions have been used to approximate the nonlinear function. In [93] and [186], different kinds of neural networks have been employed to approximate the nonlinear part of the given systems in state estimation and FD, respectively. Takagi-Sugeno (T-S) fuzzy model has been used to approximate the nonlinearities in [105, 155].

1.2.3 Communication Constraints

1.2.3.1 Transmission Delay

Due to the limited bandwidth of links and the capacity of devices, time delays occur frequently in actual communication processes. The value of communication delays is usually under the influence of various factors, including the transmission distance, the network burden, and the protocol adopted in the transmission. Denote the transmitted signal and received signal at instant k by y_k and $\hat{y}_k$, respectively, and the communication can be formulated as $\hat{y}_k = y_{k-\tau_k}$, where τ_k is the time delay.

Based on the mathematical formulation of the delay, the simplest delay is the constant delay. The sharp-cutoff finite impulse response (FIR) filters with prescribed constant group delay have been designed in [22], where a method has been proposed to determined the corresponding group delay for each subfilter by solving a set of linear equations. The parameter-dependent H_∞ filter has been established in [160] for output estimation in linear parameter varying plants with constant delays, and an LMI-based delay-dependent condition has been presented to guarantee the stability and an induced L_2 gain bound performance for the filtering error system. When the delay is constant but unknown, [202] has developed a distributed FD method for sensor faults in a formation system, where a bank of distributed fault isolation residual generators have been presented according to the closed-loop system model. With constant delays in both the states and outputs, sliding mode observers have been constructed which can guarantee H_∞ asymptotic stability of the state and fault estimation errors [123].

Naturally, the constant formulation is insufficient to describe the complex actual delays in transmission links. A widely adopted method to model the time-varying delay is the bounded description, which means that the lower and upper bounds of the delay are available to us. The delay-dependent robust H_∞ control and filtering problems for Markovian jump linear systems with norm-bounded parameter uncertainties and bounded time-varying delays have been addressed in [181], where some slack matrix variables have been introduced in the LMIs to reduce the conservatism caused by either model transformation or bounding techniques. The distributed H_∞ filtering problem has been discussed in [36] for a class of discrete-time Markovian jump nonlinear time-delay systems with deficient statistics of mode transitions. The dynamic event-triggered H_∞ filtering problem for a class of discrete time-delay complex networks with randomly occurring nonlinearities has been considered in [91], where the effects of time-varying delays have been examined in the signal transmission among the nodes. The delay-dependent H_∞ filtering design for T-S fuzzy time-varying delay systems has been studied in [39] using the input-output approach, where the H_∞ full- and reduced-order filters have both been designed in terms of LMIs. This kind of formulation is suitable to deal with both continuous and discrete time delays.

Some scholars have considered the delays as random variables and some stochastic methodologies have been employed to deal with them. Bernoulli distributed white sequences have been widely used to describe the stochastic time delays. The robust H_∞ filtering problem has been studied for a class of uncertain nonlinear networked systems with both multiple stochastic time-varying communication delays and multiple packet dropouts [34], where a linear full-order filter has been designed such that the estimation error can converge to zero exponentially in the mean square while the disturbance rejection attenuation can be constrained to a give level. The optimal least-squares state estimation problem has been addressed for a class of discrete-time multisensor linear stochastic systems with state transition and measurement random parameter matrices, correlated noises, and delayed measurements [12], where the optimal linear filter has been designed recursively using an innovation approach. In [138], a distributed EKF has been developed for a class of nonlinear systems, whose outputs have been measured by multiple sensors sending data through a communication network subject to loss and latency, and the boundedness of the filtering error has been proved under some conditions. [63] has been concerned with the optimized state estimation problem for nonlinear dynamical networks subject to random coupling strength and random sensor delays under the event-triggered communication criterion, where a new robust state estimation algorithm has been provided and a sufficient condition ensuring the boundedness of state estimation error has been presented.

One of the shortcomings to model the delay with a Bernoulli distributed white sequence is that the modeled delay is independent at each time step. Unfortunately, this is not always the case in reality. For example, the transmission burden in the link often varies slower compared with the information exchange frequency, and this means that the statistics of successive time delays may be related to each other. Markov chain has proven to be an effective way to characterize the interdependency between the delays. In [198], a T-S model has been employed to represent a networked control system with different network-induced delays and a parity-equation approach and a fuzzy-observer-based approach for fault detection have been developed. [94] has investigated the H_∞ filtering problem for a class of nonlinear stochastic systems with model uncertainties and Markov time delays, where a set of sufficient conditions have been derived for the filtering system to achieve stochastic stability in the sense of mean square and the prescribed disturbance attenuation level. [106] has been concerned with the filtering problem for a class of networked systems with Markov transmission delays and packet disordering, in which a new method has been developed to calculate the reordered transition probability matrix. [46] has studied the distributed fusion estimation for a class of multi-sensor systems with sensor gain degradation and Markovian delays, where distributed fusion estimators have been obtained as a matrix-weighted linear combination of the local filters.

There have been some other results focusing on more complex delay models. In some specific cases, the delay may appear in a random manner, rather than a deterministic one. In [132], the fusion estimation problem has been investigated for a class of multi-rate power systems with randomly occurring delays in supervisory control and data acquisition measurements. [182] has studied the issue of robust state estimation for coupled neural networks with parameter uncertainty and randomly occurring distributed delays, where the polytopic model has been employed to describe the parameter uncertainty. Another frequently encountered phenomenon is that there may be more than one type of delay in the addressed systems, and naturally, this leads to some research on the mixed delay issue. In [185], a class of discrete-time networked nonlinear systems with mixed random delays and packet dropouts has been introduced, and the H_∞ filtering problem for such systems has been investigated. A outlier-resistant state estimation problem has been addressed for a class of recurrent neural networks with mixed time delay [95], where the explicit characterization of the estimator gain has been obtained by solving a convex optimization problem.

1.2.3.2 Missing Measurement

The missing measurement phenomenon, which is also referred to as package dropout, may result from many different reasons, for example, the disconnection of a link, transmission error, information disorder, etc. The state estimation and FD problems with missing measurements have been gaining momentum in the last few decades. According to the transmitted signal is with or without the time stamp, whether the arrival of the signal is known to the receiver can be determined, the results focusing on the missing measurements can be categorized accordingly as well.

When the information on arrivals is unknown, the missing measurements have mostly been modeled with certain random sequences, and some statistical properties have been used to establish the desired filters/FD units. [25] has been concerned with a hybrid distributed dynamic state estimation algorithm for large-scale power grids, where a group of Bernoulli random processes has been used to formulate the missing measurements. Different from the centralized Kalman filter algorithm, the presented distributed approach has been capable of independently estimating local states by local measurements. In [191], the distributed recursive filtering has been considered for the discrete-time nonlinear multisensor networked system with multi-step sensor delays, missing measurements and correlated noise, where a set of Bernoulli distributed random variables has been selected to describe the multi-step sensor delays, missing measurements, and correlated noise. The filter has been designed based on linear fitting and weighted average consensus to solve the nonlinear state estimation in the multi-sensor networked system. [97] has studied the centralized fusion and weighted measurement fusion robust steady-state Kalman filtering problem for a class of multisensor networked systems with mixed uncertainties including multiplicative noises, two-step random delays, missing measurements, and uncertain noise variances, and several Bernoulli distributed sequences have been used to model the time delay and missing measurement. According to the minimax robust estimation principle and the worst-case fusion systems with conservative upper bounds of uncertain noise variances, the robust steady-state fusion Kalman estimators (predictor, filter, and smoother) have been presented in a unified framework. In [30, 177], the stabilization issue of linear systems over networks with packet loss has been thoroughly discussed, where two types of packet-loss processes have been considered. One has been the arbitrary packet-loss process, and the other has been the Markovian packet-loss process.

With the information provided by time stamps, another method to cope with the intermittent measurements is to take them as packet arrivals with known values. [147] has analyzed the convergence conditions of discrete Kalman filter with intermittent observations. It has been proved that there exists a critical value of the dropout probability below which estimation error covariance is unbounded. In the sequential study [78], the weak convergence of the estimation error covariance to a unique invariant distribution has been discussed. [107] has proposed a practical approach to realize trajectory tracking control of batch product quality in those situations where only intermittent measurements have been available. The presented recursive formulation can allow identified model to be readily incorporated as a predictor into standard

Analysis and Synthesis of Imperfect Measurements 11

model predictive control framework. In [174], the stability of Kalman filtering over Gilbert-Elliott channels with the random packet dropout described by a two-state Markov chain have been addressed. It has been proved that there exists a critical curve in the failure-recovery rate plane, below which the Kalman filter can be mean-square stable. [9] has proposed a framework for attitude estimation on the Special Orthogonal group using intermittent body-frame vector measurements, where both cases have been considered, including the case where the vector measurements are synchronously intermittent and the case where the vector measurements are asynchronously intermittent. The proposed observers have had a measurement-triggered structure where the attitude has been predicted using the continuously measured angular velocity when the vector measurements have not been available, and adequately corrected upon the arrival of the vector measurements.

1.2.3.3 Signal Quantization

As an inevitable procedure in many analog/digital conversions, signal quantization leads to unavoidable quantization errors in the transmitted signals. Two of key problems in the estimation/FD problems with quantized measurements are how to characterize the filter/FD unit properly to handle the quantization error and what effects of the error have on the desired performance. As early as 1956, Kalman has pointed out that the controlled output might be subject to some complicated behaviors such as limit circle and chaos with quantization [76]. Generally speaking, quantizers can be categorized into static quantizers and dynamic quantizers. A static quantizer can be seen as a memoryless nonlinear function mapping the signals into a quantization set. Based on the type of nonlinear functions, the static quantizer can be further classified into the uniform quantizer, the logarithmic quantizer, the probabilistic uniform quantizer, and so on.

The set of quantization levels for a uniform quantizer can be describe by

$$\mathcal{U} = \{\tau_t | \tau_t \triangleq t\Delta, \ t = 0, \pm 1, \pm 2, \cdots \}, \ \Delta > 0. \tag{1.3}$$

The quantization function $Q(.)$ maps the transmitted signal $y \in \mathbb{R}$ into the set $\mathcal{U}$. When $\tau_t \leq y < \tau_{t+1}$, the signal $Q(y) = \tau_t$.

Based on (1.3), it is clear that the quantization error for a uniform quantizer can be seen as an additive and norm-bounded disturbance. [32] has been concerned with the distributed recursive filtering problem for a class of discrete time delayed stochastic systems subject to both uniform quantization and deception attack effects on the measurement outputs. A distributed recursive filter has been designed such that an upper bound for the filtering error covariance can be guaranteed and subsequently minimized via a gradient-based method. In [167], the resilient filtering problem has been investigated for a class of linear time-varying repetitive processes with uniform quantizations and Round-Robin protocols. An upper bound has been guaranteed on the filtering error variance and subsequently minimized at each time instant, and the boundedness issue has also been discussed with respect to the filtering error variance. [101] has investigated the EKF problem for a class of stochastic nonlinear systems with quantization effects and Round-Robin communication protocols. By solving two coupled Riccati-like difference equations, the filter gain matrix has been explicitly formulated, and a sufficient condition has been established to ensure the uniform boundedness of the filtering error in the mean-square sense. [137] has focused on the linear minimum mean square estimator for a networked discrete time-varying linear system subject to uniform data quantification and communication constraints. By using orthogonal projection principle and innovation analysis method, a Kalman type filter has been designed in a recurrence form. [55] has investigated the ultimately bounded filtering problem for a kind of time-delay nonlinear stochastic systems with random access protocol and uniform quantization effects. A nonlinear filter has been devised by resorting to the stochastic analysis technique and the Lyapunov stability theory such that the filtering error dynamics can be exponentially ultimately bounded in mean square.

To guarantee the asymptotic behaviours of the networked systems with uniform quantizers [41, 71], a probabilistic uniform quantizer is introduced. A probabilistic uniform quantizer has the same quantization set as a uniform quantizer, while the quantizer function of the former is stochastic. To be more specific, in a probabilistic uniform quantizer, the quantizer output is defined as follows:

$$\begin{cases} \mathbb{P}\{\mathcal{Q}(y) = \tau_t\} = 1 - r \\ \mathbb{P}\{\mathcal{Q}(y) = \tau_{t+1}\} = r \end{cases} \qquad (1.4)$$

where $r = \frac{y - \tau_t}{\delta} \in [0, 1]$. In this case, the quantization error (i.e., $y - \mathcal{Q}(y)$) becomes an additive random variable with zero mean and bounded variance. A mean-squared error analysis and convergence of the consensus error has been proved as time goes to infinity in [15]. Furthermore, [100] has been concerned with the remote state estimation problem for a class of linear discrete time-varying non-Gaussian systems with multiplicative noises and probabilistic uniform quantizations. By introducing a proper augmented system which aggregates the original state vector and its second-order Kronecker power, the quadratic estimation problem has been transferred into a corresponding linear estimation problem of the augmented state vector. The monotonicity of the optimized upper bound with respect to the quantization accuracy has been discussed as well.

In a typical logarithmic quantizer, the set of quantization levels for the logarithmic quantizer is described by

$$\mathcal{U} = \{\tau_t | \tau_t \triangleq \pm \rho^t \tau_0, \ t = 0, \pm 1, \pm 2, \cdots\} \bigcup \{0\}. \qquad (1.5)$$

Analysis and Synthesis of Imperfect Measurements 13

where $0 < \rho < 1$, $\tau_0 > 0$. The quantized function $Q(\cdot)$ is an odd function with the following form:

$$Q(y) = \begin{cases} \tau_t, & \frac{1}{1+\delta}\tau_t \leq y \leq \frac{1}{1-\delta}\tau_t \\ 0, & y = 0 \\ -Q(-y), & y < 0 \end{cases} \quad (1.6)$$

where $\delta = (1-\rho)/(1+\rho)$. It can be seen that the quantization error for a logarithmic quantizer can be regarded as a multiplicative and norm- or sector-bounded noise.

In [142], the robust H_∞ finite-horizon filtering problem has been investigated for discrete time-varying stochastic systems with polytopic uncertainties, randomly occurred nonlinearities as well as quantization effects. A new robust H_∞ filtering technique has been developed for the addressed discrete time-varying stochastic systems, where the information of both the current measurement and the previous state estimate has been employed to estimate the current state. In [161], a novel H_∞ filtering approach has been developed for a class of discrete time-varying systems subject to missing measurements and logarithmic quantization effects. A necessary and sufficient condition has been established for the existence of the desired time-varying filters in virtue of the solvability of certain coupled recursive Riccati difference equations. [133] has proposed a novel robust EKF for a class of discrete time-varying nonlinear complex networks with event-triggered communication and logarithmic quantization effects, and the estimator parameters for each node have been derived separately into two Riccati-like difference equations so that the achieved upper bound can be minimized. [192] has examined the adaptive event-triggered fault detection problem of semi-Markovian jump systems with logarithmic output quantization, where novel sufficient conditions have been established for the stochastic stability in the proposed fault detection scheme with an H_∞ performance. In [102], the recursive filtering problem has been considered for stochastic parameter systems subject to logarithmic quantization effects and packet disorders, and sufficient conditions have been established to ensure the mean-square boundedness of filtering errors.

A dynamic quantizer can adjust the quantization strategy according to the system data and thus is more complicated but effective than a static one. In [19], the generalized H_2 filter design problems have been addressed for a class of nonlinear discrete-time systems with measurement quantization, where the system measurement output has been quantized by a dynamic quantizer constituted by a static quantizer and a dynamic parameter. Both full-and reduced-order filters and the quantizer dynamic parameter have been designed such that the quantized filtering error systems can be asymptotically stable with prescribed generalized H_2 performances. [20] has studied the robust H_∞ filtering problem for the in-vehicle networked system with sensor failure, dynamic quantization and data dropouts, where the dynamic quantizer parameter can be adjusted to realize a suitable range and quantization error and guarantee the desired system performance. The proposed filter design method has been given in the form of LMIs, guaranteeing that the filtering error system can be stochastically stable with H_∞ performance index. [194] has been concerned with the fault detection problem for a class of networked multi-rate systems with nonuniform sampling and dynamic quantization. Sampling-interval-dependent fault detection filters have been designed such that the robustness of residuals with respect to the disturbance and the sensitivity of the residuals against the fault can be guaranteed. [207] has been concerned with the moving horizon estimation (MHE) problem for networked linear systems with unknown inputs under dynamic quantization effects. A novel MHE strategy has been developed to cope with systems with unknown inputs by dedicatedly introducing certain temporary estimates of unknown inputs. The decoupling parameter of the moving horizon estimator has been designed based on certain assumptions on system parameters and quantization parameters, and the convergence parameters have been obtained such that the estimation error dynamics can be ultimately bounded. In [74], the networked filtering problem for a class of robotic manipulators with semi-Markov type parameters has been investigated under the passivity framework. In particular, a mode-dependent quantization has been proposed, in which the quantization density has been related to the current system mode. By doing so, each corresponding mode-dependent quantizer can effectively deal with the system jumping behaviours accordingly.

1.2.3.4 Sensor Gain Degradation

In engineering practice, the phenomenon of sensor degradation may occur as well, which is caused by various factors ranging from sensors aging and sensor intermittent failure to transmission congestions. In [61], the sensor gain degradation has been considered as a kind of fault, and the degradations have been described by a stochastic variable obeying the uniform distribution in a known interval. Attention has been focused on the design of a state estimator such that for all possible sensor faults and all external disturbances, the filtering error dynamic can be asymptotically mean-square stable as well as fulfil a prescribed disturbance attenuation level. [98] has been concerned with the distributed filtering problem for a class of discrete time-varying systems with stochastic nonlinearities and sensor degradation over a finite horizon. To account for the topological information of the sensor networks, a novel matrix simplification technique has been utilized to preserve the sparsity of the gain matrices in accordance with the given topology, and the analytical parameterization has been obtained for the gain matrices of the desired suboptimal filter. Furthermore, a sufficient condition has been established to guarantee the mean-square boundedness of the estimation errors. [45] has addressed the linear least-squares estimation of a signal from measurements subject to stochastic sensor gain degradation and random delays during the transmission, where the sensor gain degradation has been represented by a white sequence of random variables with values in $[0,1]$. Recursive prediction, filtering and fixed-point smoothing algorithms have been obtained using the first and second-order moments of the signal and the processes present in the observation model. [180] has focused on the fault estimation problem for a class of nonlinear systems with sensor gain degradation and stochastic protocol based on strong tracking filtering, and the phenomenon of the sensor gain degradation has been described by sequences of stochastic variables in a known interval. An augmented system has been constructed by combining the original system state vectors and the related faults into an augmented state vectors, and the strong tracking filter has been designed by introducing a fading factor into the filter structure to deal with the possible abrupt faults.

1.2.3.5 Event-Triggered Transmission

In practical engineering, neither the transmission capacities of the communication channels nor the data processing capability of the devices/components is unlimited. Consequently, the event-based communication mechanism has aroused considerable research attention due to its advantage in reducing unnecessary data transmissions and saving limited network resources. In an event-triggering strategy, the current signal is transmitted only if certain pre-specified triggering condition is satisfied, thereby effectively reducing the energy consumption.

In [117], a kind of send-on-delta (SOD) data collecting strategy has been proposed to capture information from the environment. In such a strategy, the sampling is triggered if the signal deviates by delta defined as the significant change of its value. Furthermore, the effectiveness of the method has been derived. It has been shown that the lower bound of the SOD effectiveness is independent of the sampling resolution and constitutes the built-in feature of the input signal. In this SOD framework, [121, 153] have been concerned with improving performance of a state estimation problem. When the estimator node does not receive data from the sensor node, the sensor value is in a known interval from the last transmitted sensor value. An algorithm has been proposed accordingly which can reduce the sensor value interval in certain situations, and thus the overall estimation performance can be improved without any changes in the SOD algorithms of the sensor nodes. It has been observed in [116] that a SOD algorithm with a fixed threshold cannot detect the steady state error or small state oscillation, which undesirably reduces the estimation performance. It has also been pointed out in [122] that the transmission rate of the SOD method becomes large when the sensor noise is large because of a sensor data variation. Motivated by this issue, another event-driven sampling method called area-triggered has been proposed in which sensor data are sent only when the integral of differences between the current sensor value and the last transmitted one is greater than a given threshold. Through theoretical analysis and simulation results, it has been shown that in the certain cases the proposed method can not only reduce data transmission rate but also improve estimation performance in comparison with the conventional event-driven method.

Inspired by the results above, some scholars have utilized the probability density function to estimate system states under the event-based transmission strategy. In [146], a state estimator has been developed that can successfully cope with event based measurements and attain an asymptotically bounded error-covariance matrix. A general mathematical description of event sampling has been proposed, which has been used to set up a state estimator with a hybrid update, while at synchronous instants the update has been based on knowledge that the sensor value lies within a bounded subset of the measurement space. [172] has proposed an event-based sensor data scheduler for linear systems and derived the corresponding minimum squared error estimator. A desired balance has been achieved between the sensor-to-estimator communication rate and the estimation quality by selecting an appropriate event-triggering threshold. In [17], an event-based nonlinear filter has been developed, which can be implemented using approximate nonlinear filtering algorithms including particle filtering and minimum distortion filters. In [145], an optimal sensor fusion problem has been considered based on the information from multiple sensors that provide their measurement updates according to separate event-triggering conditions. It has been shown that under a commonly-accepted Gaussian assumption, the optimal estimator depends on the conditional mean and covariance of the measurement innovations. When each channel of the sensors has its own event-triggering condition, closed-form representations have been derived for the optimal estimate and the corresponding error covariance matrix, and the exploration of the set-valued information provided by the event-triggering sets can guarantee the improvement of estimation performance. [18] has addressed an event-based state estimation for a linear discrete-time system with a multiplicative measurement noise. An innovation-based scheduling scheme has been used to communicate measurements from the sensor to the remote estimator. A maximum posteriori estimator has been employed to deal with multiplicative noise, enabling the recursive equations for the updated state and its covariance.

In the minimum-variance sense, the bound of the error induced by event-triggered transmission has been widely used to construct an upper bound of the estimation error covariance. In [99], the distributed filtering problem has been investigated for a class of discrete time-varying systems with an event-based communication mechanism. In terms of an event indicator variable, the triggering information has been utilized to reduce the conservatism in the filter analysis. An upper bound for the filtering error covariance has been obtained and minimized in the form of Riccati-like difference equations, where a novel matrix simplification technique has been developed to handle the sparseness of the sensor network topology. [157] has been concerned with the event-triggered robust fusion estimation problem for uncertain multi-rate sampled-data systems with stochastic nonlinearities and the coloured measurement noises. A set of local event-triggered filters has been constructed and the upper bounds of the local filtering error covariances at each sampling instant have been obtained. A new fusion estimation scheme has been proposed with the help of covariance intersection (CI) method and the consistency of the proposed CI-based fusion estimation scheme has been shown. [64] has been concerned with the recursive filtering problem for a class of time-varying nonlinear stochastic systems in the presence of event-triggered transmissions and multiple missing measurements with uncertain missing probabilities. An upper bound of the filtering error covariance has been obtained and then minimized by properly designing the filter gain. [72] has proposed a new method for robust fault diagnosis based on data-sending management in discrete-time linear stochastic systems. A methodology has been presented for identifying a subspace independent of disturbances, and sufficient conditions have been proposed to guarantee the observability and stability of the new subspace. Event-triggered and self-triggered hybrid methods have been used to decrease the transfer amount of measured data, and at the same time to preserve the fault diagnosis method performance. To perform the nonlinear remote state estimation in the wireless sensor network with communication constraints, [90] has proposed a novel filter based on the event-triggered schedule and the cubature Kalman filter. To deal with the non-Gaussian property due to the nonlinear transformation, the third-degree spherical-radial cubature rule has been adopted. The stochastic stability of the designed filter has been analyzed considering the relationships among the estimation performance, the communication rate and the designed parameter.

Another widely adopted way to deal with the event-triggered transmission is the robust methodology. The difference between the transmitted signal and the actual measurement can be constrained by some linear conditions, and the LMI-based or some other optimization algorithms can be applied accordingly. In [33], a general event-triggered framework has been developed to deal with the finite-horizon H_∞ filtering problem for discrete time-varying systems with fading channels, randomly occurring nonlinearities, and multiplicative noises. An event indicator variable has been constructed and the corresponding event-triggered scheme has been proposed. A set of time-varying filters has been designed such that the influences from the exogenous disturbances onto the filtering errors are attenuated at the given level quantified by a H_∞-norm in the mean-square sense. [86] has investigated the problem of fault detection for nonlinear discrete-time networked systems under an event-triggered scheme. A novel polynomial event-triggered scheme has been proposed to determine the transmission of the signal. A fault detection filter has been designed to guarantee that the residual system is asymptotically stable and satisfies the desired performance, and polynomial approximated membership functions obtained by Taylor series have been employed for filtering analysis. [151] has been devoted to even-triggered reliable H_∞ filter design for a class of nonlinear partial differential equation (PDE) systems with Markovian jumping sensor faults. Time delays and signal quantization have been taken into account, and an integral-type event-triggered scheme has been developed to improve the transmission channel utilization. Non-parallel-distributed-compensation technique has been introduced to increase the flexibility of filter design and the filter parameters can be obtained by solving several LMIs. [104] has been concerned with finite-time filtering problem for networked state-dependent uncertain systems with event-triggered mechanism and multiple attacks, which consists of deception attacks, denial-of-service attacks, and replay attacks. Sufficient conditions guaranteeing the exponentially mean-square finite-time boundedness of filtering error systems have been obtained, and the explicit expression has been derived for the parameters of the desired finite-time filter. Under the discrete event-triggered communication scheme, the problems of the robust L_1 filtering for a class of uncertain networked control systems with multiple sensor fault modes and persistent amplitude-bounded disturbance constraints have been investigated in [128]. Results on stability and robust L_1 performance have been proposed for the filtering error system according to Lyapunov theory and the integral inequality method, and the co-design method for gaining the desired L_1 filter parameters and event-triggering parameters has been given in terms of LMIs.

It is also worth mentioning that some relatively novel event-triggered transmission strategies have been proposed recently to further effectively use the bandwidth of the addressed links. One of the most representative strategies is the dynamic event-triggered transmission. [53] has presented the dynamic event-triggered transmission by the introduction of an internal dynamic variable. The stability of the resulting closed-loop system has been proved and the influence of design parameters on the decay rate of the Lyapunov function has also been discussed. For linear systems, a lower bound on the inter-execution time as a function of the parameters has been obtained and the influence of these parameters on a quadratic integral performance index has also been studied. Such a notion has been extended to stochastic systems [169] and discrete-time systems [68]. [166] has dealt with the distributed state estimation problem for an array of discrete time-varying systems over sensor networks under dynamic event-based transmission scheme, random parameter matrices, and dynamic measurement quantization. An upper bound has been guaranteed on the estimation error covariance, and such an upper bound has been minimized at each time-step by choosing proper gain matrices. To overcome the difficulties induced by the sparseness of the network topology, a matrix simplification technique has been proposed. Moreover, a sufficient condition has been provided to ensure that the estimation error is bounded in the mean-square sense. [40] has been concerned with the issue of dynamic event-based non-fragile dissipative state estimation for a type of stochastic complex networks subject to a randomly varying coupling as well as fading measurements, where the variation of coupling is governed by a Markov chain. A dynamic event-based non-fragile estimator has been developed such that, for all possible parameter fluctuations in estimator gains, the estimation error system is stochastically stable. Through intensive stochastic analysis, sufficient conditions have been derived in terms of LMI to guarantee the existence of the desired state estimator.

1.2.3.6 Sensor Saturation

In many real-world applications of networked systems, it is ubiquitous that the physical sensor cannot produce measurement signals with unlimited amplitude mainly due to the hardware restrictions, and this is referred to as the sensor saturation phenomenon. Such a phenomenon, if not properly considered in the stage of filter design, is likely to result in performance degradation or even algorithm divergence. For a vector $r = [r_1, \ldots, r_m]^T$, the saturation function $\sigma : \mathbb{R}^m \to \mathbb{R}^m$ is defined as:

$$\sigma(r) = [\sigma_1(r_1), \ldots, \sigma_m(r_m)]^T \tag{1.7}$$

where $\sigma_s(r_s) = \text{sign}(r_s)\min(b_s, |r_s|)$ and $b_s \geq 0$ for all $s = 1, \ldots, m$. Furthermore, $\text{sign}(\cdot)$ denotes the signum function and b_s represents the saturation level. It is noted that the system state, control input, and measurement output in real systems may all be subject to saturations and it is the sensor saturation that is mainly discussed in the book. So far, a large amount of research attention has been devoted to the state estimation/FD problem with sensor saturations.

When a robust criterion is considered, the error brought in by the sensor saturation can be formulated as an error bounded by some linear constraints such as Lipschitz conditions and sector-bounded conditions. [35] has addressed the robust H_∞ filtering problem for a class of discrete time-varying Markovian jump systems with randomly occurring nonlinearities and sensor saturation, and a sector-bounded assumption has been employed to describe the saturation. A robust filter over a given finite horizon has been designed such that the H_∞ disturbance attenuation level is guaranteed, and sufficient conditions have been established for the existence of the desired filter. [154] has been concerned with the problem of H_∞ filtering for a class of time delay nonlinear Hamiltonian systems with wireless network communication, in which the saturation has been constrained by a sector-bounded condition. Some sufficient conditions have been proposed to obtain effective filter gain and achieve the H_∞ performance for the augmented system consisting of the Hamiltonian system and the filter. [204] has addressed the problem of event-triggered H_∞ filter design for nonlinear systems under both hybrid attacks and sensor saturation. The saturation-induced error has been handled with a Lipschitz-like condition. By using Lyapunov-Krasovskii stability theory, sufficient conditions have been obtained to guarantee the stability of the fuzzy filtering error system. [92] has studied the H_∞ fault estimation problem for linear discrete time-variant systems with actuator and sensor saturations. To handle the saturation nonlinearities, a pair of an auxiliary linear model and an associated performance function augmenting from the conventional H_∞ performance index have been constructed, and based on this pair, the original fault estimation problem has been readdressed as a two-step optimization issue with an indefinite quadratic cost function. A Krein-space based inner product interpretation for the indefinite cost function has been established and optimal linear estimation technique has been employed to derive the stationary point of the aforementioned indefinite quadratic cost function. The H_∞ filtering problem for a class of networked nonlinear Markovian jump systems subject to randomly occurring distributed delays, nonlinearities, quantization effects, missing measurements, and sensor saturation has been investigated in [124], where the measurement saturation has been described by sector nonlinearities. A quantized resilient filter has been designed that guarantees not only the stochastic stability of the augmented filtering error system but also a prespecified level of H_∞ performance.

In the minimum-variance sense, the saturation level can be used to constitute an upper bound of the estimation error covariance since the amplitude of the difference between two saturated signals is bounded. [65] has been concerned with the joint state and fault estimation problem for a class of uncertain time varying nonlinear stochastic systems with randomly occurring faults and sensor saturations. The signum function has been employed to describe the sensor saturation owing to physical limits on the measurement output. A locally optimal time-varying estimator has been designed to simultaneously estimate both the system states and the fault signals such that, at each sampling instant, the covariance of the estimation error has an upper bound that can be minimized by properly designing the estimator gain. A sufficient condition has been given to verify the exponential boundedness of the estimation error in the mean square sense. In [190], the event-triggered filtering and intermittent fault detection problems have been investigated for a class of time-varying systems with stochastic parameter uncertainty and sensor saturation. By utilizing the inductive and stochastic analysis technique, the filter gain has been designed to ensure that the upper bound of the estimation error covariance is minimized at each time step. Based on the proposed filter, a residual has been generated and the corresponding evaluation function and detection threshold have been given to achieve fault detection. In [108], an unscented Kalman filtering problem has been studied for a nonlinear system with sensor saturation and randomly occurring false data injection attacks. A modified UKF has been designed by minimizing an upper bound of filtering error covariance, and a sufficient condition has been provided to ensure an exponentially bounded filtering error in the mean square sense. [69] has been concerned with the event-triggered state and fault estimation issue for nonlinear systems with sensor saturations and fault signals. An event-triggered recursive state and fault estimator has been designed such that the estimation error covariances for the state and fault are both guaranteed with upper bounds and subsequently the gain matrices have been derived to minimize such upper bounds, relying on the solutions to a set of difference equations.

It is also worth mentioning that sometimes the sensor saturation may occur in a random manner, rather than a deterministic one, due to various reasons such as random sensor failures and abrupt environmental changes. [37] has investigated the stochastic finite-time distributed H_∞ filtering problem for more general T-S fuzzy systems with immeasurable premise variables over wireless sensor networks with switching topology. Sensor saturation and measurement missing have both been modelled by mutually independent Bernoulli processes with uncertain probability. Utilizing non-parallel-distributed-compensation scheme, a switching-type distributed filter based on estimated premise variables has been designed to realize the sharing of filtering information and measurement information. A distributed robust filtering method has been proposed to analyze the distributed filtering error system containing unknown premise variables. [144] has studied the H_∞ filtering problem for a class of discrete networked multi-rate multi-sensor systems with randomly occurring sensor saturations under the p-persistent carrier sense multiple access protocol, where a set of mutually independent Bernoulli distributed white sequences has been introduced to characterize the random occurrence of the sensor saturations. Sufficient conditions have been established on the existence of the desired H_∞ filters and the corresponding filter gains have been then characterized by resorting to the feasibility of certain matrix inequalities. [130] has investigated the $L_2 - L_\infty$ filtering problem for a class of stochastic systems with time delay and randomly occurring nonlinearities and sensor saturation. A novel stochastic time-delay system model has been established, in which randomly occurring nonlinearities and sensor saturation phenomena have been considered. A novel functional containing negative definite terms has been constructed to relax the constraints, and a new free-matrix-based stochastic integral inequality has also also given. Meanwhile, a novel $L_2 - L_\infty$ performance analysis method making full use of delay information has been proposed. [88] has been concerned with the distributed filtering problem for a class of delayed nonlinear systems with random sensor saturation under a dynamic event-triggered mechanism. Both the Bernoulli distributed random variables and saturation function have been employed to model the phenomenon of random sensor saturation. A suboptimal filter has been designed such that the covariance of the filtering error has an upper bound, which has been minimized by appropriately computing the filter gain. Furthermore, a sufficient criterion has been presented to ensure that the filtering error is mean-square bounded. In [150], the particle filtering problem has been investigated for a class of nonlinear/non-Gaussian systems with energy harvesting sensors subject to randomly occurring sensor saturations, where the random occurrences of the sensor saturations have been characterized by a series of Bernoulli distributed stochastic variables with known probability distributions. The effects of the randomly occurring sensor saturations and the possible measurement losses induced by insufficient energies have been fully considered in the design of filtering scheme, and an explicit expression of the likelihood function has been derived.

1.2.3.7 Integral Measurement

The integral measurement issue, which is frequently encountered when optic, chemical, and nuclear signals are handled, has received some initial research attention. In [54], a novel model has been put forward to describe the integral measurement, and a variable dimension unscented Kalman filter (VD-UKF) has been proposed to estimate the states based on the new model. Furthermore, the stability of the proposed VD-UKF has been analyzed. Compared with the existing results, the proposed stability condition has been significantly relaxed and the invertibility condition of Jacobian matrices has been no longer needed. [26] has handled the event-triggered distributed filtering problem for a class of discrete-time systems with integral measurements over sensor networks. A sufficient condition has been established ensuring the existence of the distributed filter under the desirable performance index. Moreover, the ultimate bound in the performance index has been minimized and the filter gains have been acquired by solving an LMI with the YALMIP toolbox. In [129], the optimal filtering for state space model with time-integral measurement has been studied in the minimum-variance sense as well.

The integral measurement has been considered in some faulty cases as well. [48] has been concerned with the Tobit Kalman filtering problem for a class of discrete time-varying systems subject to censored observations, integral measurements and probabilistic sensor failures under the Round-Robin protocol. By resorting to the augmentation technique and the orthogonality projection principle, a protocol-based Tobit Kalman filter (TKF) has been developed, and the performance of the proposed filter has been analyzed through examining the statistical property of the error covariance of the state estimation. Further analysis has shown the existence of self-propagating upper and lower bounds on the estimation error covariance. [206] has discussed the parity space-based fault detection method for a class of linear discrete-time systems with integral measurements. A novel parity relation has been established to tackle integral measurements, and the parameters of the fault detection unit have been redesigned such that the generated residual signal is simultaneously decoupled from initial states, robust against disturbances, and sensitive to the faults. The singular value decomposition algorithm has been employed to calculate parity space matrices.

1.3 Outline of This Book

This book is divided into eleven chapters, where Chapter 2–Chapter 5 mainly focus on the state estimation problems, and Chapter 6–Chapter 10 are concerned with the FD problems. To be specific, the outline of this book is given as follows:

Outline of This Book 23

- Chapter 1 presents some concepts and challenges in imperfect measurements, reviews research progress of filtering and FD problems with disturbances, nonlinearities, and transmission constraints, and lists the outline of the book.

- Chapter 2 investigates the optimal filtering problem for a class of networked systems in the presence of stochastic sensor gain degradations. The degradations are described by sequences of random variables with known statistics. A new measurement model is put forward to account for sensor gain degradations, network-induced time delays as well as network-induced data dropouts. Based on the proposed new model, an optimal unbiased filter is designed that minimizes the filtering error variance at each time-step. The developed filtering algorithm is recursive and therefore suitable for online application. Moreover, both currently and previously received signals are utilized to estimate the current state in order to achieve a better accuracy.

- Chapter 3 is concerned with the minimum variance filtering problem for a class of time-varying systems with both additive and multiplicative stochastic noises through a sensor network with a given topology. The measurements collected via the sensor network are subject to stochastic sensor gain degradation, and the gain degradation phenomenon for each individual sensor occurs in a random way governed by a random variable distributed over the interval $[0,1]$. The purpose of the addressed problem is to design a distributed filter for each sensor such that the overall estimation error variance is minimized at each time step via a novel recursive algorithm. By solving a set of Riccati-like matrix equations, the parameters of the desired filters are calculated recursively. The performance of the designed filters is analyzed in terms of the boundedness and monotonicity. Specifically, sufficient conditions are obtained under which the estimation error is exponentially bounded in mean square. Moreover, the monotonicity property for the error variance with respect to the sensor gain degradation is thoroughly discussed.

- Chapter 4 discusses the filtering problem for a class of nonlinear systems with stochastic sensor saturations and Markovian measurement transmission delays, where the asymptotic stability in probability is considered. The sensors are subject to random saturations characterized by a Bernoulli distributed sequence. The transmission time delays are governed by a discrete-time Markov chain with finite states. In the presence of the nonlinearities, stochastic sensor saturations, and Markovian time delays, sufficient conditions are established to guarantee that the filtering process is asymptotically stable in probability without disturbances and also satisfies the H_∞ criterion with respect to non-zero exogenous disturbances under the zero-initial condition. Moreover, it is illustrated that the results can be specialized to linear filters.

- Chapter 5 deals with the observer design problem for a class of systems with unknown inputs and missing measurements. A Bernoulli distributed sequence taking values on 0 or 1 is employed to govern possible missing measurements. It is worth mentioning that the effects of the unknown inputs cannot be fully eliminated from the estimation error due to the stochastic nature of the addressed system. In such a case, an observer is established which can decouple the estimation error from the unknown inputs in the mathematical expectation sense. Moreover, the uniform ultimate boundedness of the estimation error is discussed.

- Chapter 6 is concerned with polynomial filtering and fault detection problems for a class of nonlinear systems subject to additive noises and faults. The nonlinear functions are approximated with polynomials of a chosen degree. Different from the traditional methods, the approximation errors are not discarded but formulated as low-order polynomial terms with norm-bounded coefficients. The aim of the filtering problem is to design a least squares filter for the formulated nonlinear system with uncertain polynomials, and an upper bound of the filtering error covariance is found and subsequently minimized at each time step. The desired filter gain is obtained by recursively solving a set of Riccati-like matrix equations, and the filter design algorithm is therefore applicable for online computation. Based on the established filter design scheme, the fault detection problem is further investigated where the main focus is on the determination of the threshold on the residual. Due to the nonlinear and time-varying nature of the system under consideration, a novel threshold is determined that accounts for the noise intensity and the approximation errors, and sufficient conditions are established to guarantee the fault detectability for the proposed fault detection scheme.

- Chapter 7 is concerned with the filtering problem for a class of nonlinear systems with stochastic sensor saturations and event-triggered measurement transmissions. An event-triggered transmission scheme is proposed with hope to ease the traffic burden and improve the energy efficiency. The measurements are subject to randomly occurring sensor saturations governed by Bernoulli-distributed sequences. Special effort is made to obtain an upper bound of the filtering error covariance in the presence of linearization errors, stochastic sensor saturations as well as event-triggered transmissions. A filter is designed to minimize the obtained upper bound at each time step by solving two sets of Riccati-like matrix equations, and thus the recursive algorithm is suitable for online computation. Sufficient conditions are established under which the filtering error is exponentially bounded in mean square. The applicability of the presented method is demonstrated by dealing with the fault estimation problem.

Outline of This Book 25

- Chapter 8 investigates the finite-horizon quantized H_∞ filter design problem for a class of time-varying systems with quantization effects and event-triggered measurement transmissions. A componentwise event-triggered transmission strategy is put forward to reduce the unnecessary communication burden for the purpose of energy efficiency. The transmitted measurements triggered according to prespecified events are quantized by a logarithmic quantizer. Special attention is paid to the design of the filter such that a prescribed H_∞ performance can be guaranteed over a given finite horizon in the presence of nonlinearities, quantization effects and event-triggered transmissions. Two sets of Riccati difference equations are introduced to ensure the H_∞ estimation performance of the designed filter. The filter design algorithm is recursive and thus suitable for online computation.

- Chapter 9 deals with a fundamental issue of evaluating the performance of widely used fault detection and diagnosis (FDD) schemes within a closed-loop framework. The focus is to examine how certain implemented controller would impact on the FDD performance and how such performance can be further improved. For this purpose, the authors consider the FDD problem for a class of linear discrete-time systems (with and without unknown disturbances) under typical proportional-integral control using observer-based methods. It is revealed that some existing observer-based FDD approaches are no longer applicable in the closed-loop situation due to the feedback control. To solve the problem, by appropriately modifying the structure and redesigning the parameters of the observers, it is proven that the dynamics of closed-loop residuals can be made identical with those of the residuals obtained with known control inputs at each time step.

- Chapter 10 is concerned with the state estimation and fault reconstruction problems for a class of discrete systems with integral measurements under partially decoupled disturbances. The considered integral measurements, as functions of the system states over a period of time, reflect the interval time between sample collections and real-time signal processing. Moreover, the process disturbances are allowed to be partially decoupled in the observer design. An augmented state vector is constructed, which consists of the current system state, the delayed system state and the additive fault, and the resultant augmented system is described in a singular form. Then, an unknown input observer is obtained that decouples partial disturbances and attenuates the effect from the remaining undecouplable disturbances. The existence conditions of the desired observer are thoroughly investigated and an algorithm for designing the observer gains is also provided.

- Chapter 11 draws some conclusions on the book, and points out some potential research directions related to the work done in this book.

In addition, the illustrative numerical simulation examples, which are employed to validate the effectiveness of the results, are provided within the individual chapters.

2

Optimal Filtering for Networked Systems with Stochastic Sensor Gain Degradation

In the past decade, the research on networked systems has been gaining momentum due to the rapid advances in communication technology and the increasing interest in systems connected via wireless and shared links. In networked systems, it is often the case that the system outputs are measured by various sensors spatially located in a wide area, where the measurements are transmitted to a remote central estimator parallelly. Measurements in a networked environment might be seriously impaired by imperfect data transmissions, for example, communication delay and packet dropout, which may deteriorate the performance or cause instability. Thus, many traditional control/filtering techniques, which focus on interconnected dynamical systems linked through ideal channels, should be redesigned before being applied in networked systems. Much research attention has been devoted to control/filtering in networked systems to deal with the phenomena that shared communication network induces.

However, the sensor gain degradation issue in a networked environment has not been investigated extensively. Sensor gain variation occurs frequently in engineering practice. This is particularly true for real-world systems under changeable working conditions. Examples include thermal sensors for vehicles [183], radiation detectors for nuclear/radiological threat [112], and the platform mounted sonar for the acoustic signals from the ocean [148]. It is quite common in a networked system that the sensor gains degrade in a random fashion, which is due probably to sensor aging, intermittent sensor outages, or network-induced saturations/congestions. As such, the so-called stochastic sensor gain degradation has received some initial research attention [59]. To date, the corresponding results on *time-varying* systems using *recursive algorithms* under *stochastic sensor gain degradations* have been very few in the literature. Moreover, the network-induced time delay and data dropout during transmissions over shared networks should be taken into consideration.

In this chapter, we are motivated to study the optimal filtering problem for networked time-varying systems with stochastic sensor gain degradations by a recursive matrix equation approach. In the developed filtering algorithm, both current and previous measurements are used to estimate the system state so as to achieve a better estimation performance. The filter parameters are calculated recursively and the developed algorithm is thus suitable for online computation. The main contributions of the chapter are outlined as follows: 1) a new model is proposed that is comprehensive to cater for sensor gain degradations, network-induced time delays as well as data dropouts within a unified framework; 2) an optimal unbiased filter is designed that minimizes the filtering error variance under sensor gain degradations whose occurrence probability is allowed to vary over the interval $[0,1]$; and 3) the developed recursive algorithm is shown to be both effective and efficient as compared with the standard Kalman filtering method.

2.1 Problem Formulation and Preliminaries

Consider the following class of time-varying discrete-time linear systems:

$$x_{k+1} = \left(A_k + g_k \hat{A}_k\right) x_k + w_k, \qquad (2.1)$$

where the matrices A_k and $\hat{A}_k$ are known; $x_k \in \mathbb{R}^n$ is the state; $w_k \in \mathbb{R}^n$ is the additive white noise with $\mathbb{E}\{w_k\} = 0$ and $\mathbb{E}\{w_k w_k^T\} = W_k$; g_k is the multiplicative white noise with $\mathbb{E}\{g_k\} = \bar{g}_k$ and $\mathbb{E}\{g_k^2\} = \tilde{g}_k$. Assuming that there are μ different classes of sensors and the sensors belonging to the same class receive/send signals in one go, the measurements before transmitting are described by

$$y_k^{(i)} = f_k^{(i)} C_k^{(i)} x_k + v_k^{(i)}, i = 1, 2, \ldots, \mu, \qquad (2.2)$$

where $C_k^{(i)}$ is a known matrix; $y_k^{(i)} \in \mathbb{R}^{m^{(i)}}$ is the measurement of the sensors of class i ; $v_k^{(i)} \in \mathbb{R}^{m^{(i)}}$ is the measurement noise of the sensors of class i with $\mathbb{E}\left\{v_k^{(i)}\right\} = 0$ and $\mathbb{E}\left\{v_k^{(i)}\left(v_k^{(i)}\right)^T\right\} = V_k^{(i)}$. $v_k^{(i)}$ is assumed to be independent of w_k and g_k. $f_k^{(i)}$ is a random variable distributed over the interval $[a^{(i)}, b^{(i)}]$ $\left(0 \leq a^{(i)} \leq b^{(i)} \leq 1\right)$ with $\mathbb{E}\left\{f_k^{(i)}\right\} = \bar{f}_k^{(i)}$ and $\mathbb{E}\left\{\left(f_k^{(i)}\right)^2\right\} = \tilde{f}_k^{(i)}$, where $\bar{f}_k^{(i)}$ and $\tilde{f}_k^{(i)}$ are known scalars.

Remark 2.1 *The random variable $f_k^{(i)}$ accounts for the stochastic sensor gain degradation. Compared with existing literature, (2.2) provides a more precise means for quantifying the sensor gain degradations, since $f_k^{(i)}$ is not restricted to take values at 0 or 1 only, and its statistical properties could be time-varying. When $f_k^{(i)}$ specializes to the traditional Bernoulli distributed one, our model covers the binary one in [168].*

Problem Formulation and Preliminaries

The received signal impaired by communication delays and data dropouts can be described as:

$$\begin{cases} r_k^{(i)} = \sum_{j=0}^{L} \delta\left(\tau_k^{(i)}, j\right) y_{k-j}^{(i)} + n_k^{(i)}, \\ y_d^{(i)} = 0, d = -L, -L+1, \ldots, -1, \end{cases} \quad (2.3)$$

where L is the maximum time delay; $r_k^{(i)} \in \mathbb{R}^{m^{(i)}}$ is the received signal of the ith channel; $n_k^{(i)} \in \mathbb{R}^{m^{(i)}}$ is the transmission white noise of the ith channel with $\mathbb{E}\left\{n_k^{(i)}\right\} = 0$ and $\mathbb{E}\left\{n_k^{(i)}\left(n_k^{(i)}\right)^T\right\} = N_k^{(i)}$, and is independent of w_k and $v_k^{(i)}$. $\delta(\cdot)$ is the standard Dirac function with $\mathbb{E}\left\{\delta\left(\tau_k^{(i)}, j\right)\right\} = \text{Prob}\left\{\tau_k^{(i)} = j\right\} = p_{j,k}^{(i)}$ and $\sum_{j=0}^{L} p_{j,k}^{(i)} \leq 1$. When $\tau_k^{(i)} = j$, the time delay is j at time k at the sensors of class i. When $\tau_k^{(i)} = 0$, the transmission is perfect, and no time delay or data dropout occurs. During the transmission at time k, the measurements get lost with the probability $1 - \sum_{j=0}^{L} p_{j,k}^{(i)}$. With knowledge on the order of the received signal, the receiver would discard the previously transmitted packets when several measurements arrive at the receiver in the same interval.

The following assumptions are needed in the derivation of the main results.

Assumption 2.1 *For any i and k, $f_k^{(i)}$, g_k and $\tau_k^{(i)}$ are independent of each other and all the noises.*

Assumption 2.2 *The initial state x_0 is stochastic with $\mathbb{E}\{x_0 x_0^T\} = X_{0,0}$, and x_0 is independent of the noises, transmissions, and sensor gain degradations.*

For notational convenience, we denote

$$\hat{x}_0 = \mathbb{E}\{x_0\},$$

$$r_k = \left[\left(r_k^{(1)}\right)^T, \ldots, \left(r_k^{(\mu)}\right)^T\right]^T,$$

$$\bar{F}_k = \text{diag}\left\{\bar{f}_k^{(1)} I_{m^{(1)}}, \ldots, \bar{f}_k^{(\mu)} I_{m^{(\mu)}}\right\},$$

$$\bar{P}_{j,k} = \text{diag}\left\{p_{j,k}^{(1)} I_{m^{(1)}}, \ldots, p_{j,k}^{(\mu)} I_{m^{(\mu)}}\right\},$$

$$\tilde{C}_k = \left[\left(C_k^{(1)}\right)^T, \ldots, \left(C_k^{(\mu)}\right)^T\right]^T.$$

In this chapter, we are interested in designing a full-order filter of the following form

$$\hat{x}_{k+1} = L_k \left(r_k - \sum_{j=0}^{L} \bar{P}_{j,k} \bar{F}_{k-j} \tilde{C}_{k-j} \hat{x}_{k-j}\right) + \left(A_k + \bar{g}_k \hat{A}_k\right) \hat{x}_k, \quad (2.4)$$

where L_k are filter parameters to be determined.

The filter structure given in (2.4) is set so as to achieve the unbiased estimation. The unbiasedness can be proven by mathematical induction as follows.

Denote $\tilde{x}_k = x_k - \hat{x}_k$. Firstly, one can verify that the assertion is true for $k = 0$ according to $\hat{x}_0 = \mathbb{E}\{x_0\}$. Secondly, we assume that it is true for the integers from 0 to k, then we have

$$\mathbb{E}\{\tilde{x}_{k+1}\}$$
$$= \mathbb{E}\left\{\left(A_k + g_k\hat{A}_k\right)x_k + w_k\right\} - \left(A_k + \bar{g}_k\hat{A}_k\right)\hat{x}_k$$
$$- L_k\left(\mathbb{E}\{r_k\} - \sum_{j=0}^{L}\bar{P}_{j,k}\bar{F}_{k-j}\tilde{C}_{k-j}\hat{x}_{k-j}\right)$$
$$= \left(A_k + \bar{g}_k\hat{A}_k\right)\mathbb{E}\{\tilde{x}_k\} - L_k\sum_{j=0}^{L}\bar{P}_{j,k}\bar{F}_{k-j}\tilde{C}_{k-j}\mathbb{E}\{\tilde{x}_{k-j}\}$$
$$= 0.$$

This concludes the proof.

Remark 2.2 *A particular feature of the structure in (2.4) is that both currently and previously received signals are utilized to estimate the state of the system. It is worth mentioning that the state augmentation method has been largely used in the literature to tackle the measurement delays, which would inevitably cause exponentially increasing computational burden due to the increased dimension. In contrast, the recursive matrix equation method to be developed in this chapter possesses the advantage of efficient computation for the addressed filter design problem. On the other hand, note that the terms reflecting the statistics, $\bar{F}_k$, $\bar{P}_{j,k}$ and $\bar{g}_k$, are fixed values which make the filter structure easy-to-implement.*

Our goal in this chapter is to design the filter parameter L_k that can achieve the minimum filtering error variance based on the received signals $\{r_0, \ldots, r_{k-1}, r_k\}$ at each time step k. Note that the exact value of the random time delay is not required to be measured at the receiver for each time step. Instead, only the information about the statistical law of the occurrence of the time delays is needed, where the statistical law could be obtained through round-trip-time-test [200]. In this sense, the proposed filter has certain robustness against the random changes and possible measurement errors of the delays.

2.2 Optimal Filter Design

Before proceeding, we define the following notations: **1** represents matrix whose entries are all one; * in a symmetric block matrix represents a term

Optimal Filter Design 31

that can be determined by symmetry; ∘ is the Hadamard product with this product being defined as $[A \circ B]_{ij} = A_{ij}B_{ij}$.

Theorem 2.1 *The parameter of the filter (2.4) achieving the minimum filtering error variance is given by:*

$$L_k = S_k^T T_k^{-1}, \qquad (2.5)$$

where

$$V_k = \mathrm{diag}\left\{V_k^{(1)}, \ldots, V_k^{(\mu)}\right\},$$

$$N_k = \mathrm{diag}\left\{N_k^{(1)}, \ldots, N_k^{(\mu)}\right\},$$

$$X_{i,j} = \begin{cases} A_{i-1}X_{i-1,i-1}A_{i-1}^T + W_{i-1} + \bar{g}_{i-1}\left(\hat{A}_{i-1}X_{i-1,i-1}A_{i-1}^T \right. \\ \left. + A_{i-1}X_{i-1,i-1}\hat{A}_{i-1}^T\right) + \tilde{g}_{i-1}\hat{A}_{i-1}X_{i-1,i-1}\hat{A}_{i-1}^T, & \text{if } i = j, \\ \prod_{m=1}^{i-j}\left(A_{i-m} + \bar{g}_{i-m}\hat{A}_{i-m}\right)X_{j,j}, & \text{if } i > j, \\ X_{i,i}\prod_{m=1}^{j-i}\left(A_{j-m} + \bar{g}_{j-m}\hat{A}_{j-m}\right)^T, & \text{if } i < j, \end{cases}$$

$$\tilde{F}_k = \begin{bmatrix} \tilde{f}_k^{(1)}\mathbf{1}_{m^{(1)}\times m^{(1)}} & * & * & & * \\ \bar{f}_k^{(2)}\bar{f}_k^{(1)}\mathbf{1}_{m^{(2)}\times m^{(1)}} & \ddots & * & & * \\ \vdots & & \vdots & \ddots & \\ & & & & * \\ \bar{f}_k^{(\mu)}\bar{f}_k^{(1)}\mathbf{1}_{m^{(\mu)}\times m^{(1)}} & \cdots & \cdots & \tilde{f}_k^{(\mu)}\mathbf{1}_{m^{(\mu)}\times m^{(\mu)}} \end{bmatrix},$$

$$Q_{j,k} = \begin{bmatrix} p_{j,k}^{(1)}\mathbf{1}_{m^{(1)}\times m^{(1)}} & * & * & & * \\ p_{j,k}^{(2)}p_{j,k}^{(1)}\mathbf{1}_{m^{(2)}\times m^{(1)}} & \ddots & * & & * \\ \vdots & & \vdots & \ddots & \\ & & & & * \\ p_{j,k}^{(\mu)}p_{j,k}^{(1)}\mathbf{1}_{m^{(\mu)}\times m^{(1)}} & \cdots & \cdots & p_{j,k}^{(\mu)}\mathbf{1}_{m^{(\mu)}\times m^{(\mu)}} \end{bmatrix},$$

$$Y_{i,j} = \begin{cases} \tilde{F}_i \circ \tilde{C}_i X_{i,i} \tilde{C}_i^T + V_i, & \text{if } i = j, \\ \bar{F}_i \tilde{C}_i X_{i,j} \tilde{C}_j^T \bar{F}_j^T, & \text{if } i \neq j, \end{cases}$$

$$R_{i,j} = \begin{cases} \sum_{m=0}^{L} Q_{m,i} \circ Y_{i-m,i-m} + N_i, & \text{if } i = j, \\ \sum_{m=0}^{L}\sum_{n=0}^{L} \bar{P}_{m,i} Y_{i-m,j-n} \bar{P}_{n,j}^T, & \text{if } i \neq j, \end{cases}$$

$$H_{i,j} = m_j X_{i,j} C_j^T,$$

$$M_{i,j} = \sum_{m=0}^{L} p_{m,j} H_{i,j-m},$$

$$U_{i,j} = U_{i,j-1}\left(A_{j-1} + \bar{g}_{j-1}\hat{A}_{j-1}\right)^T + \left(R_{i,j-1} - \sum_{l=0}^{L} U_{i,j-1-l}\tilde{C}_{j-1-l}^T \bar{F}_{j-1-l}^T\right.$$
$$\left. \times \bar{P}_{j-1-l}^T \right) L_{j-1}^T,$$

$$\Lambda_{i,j} = \Lambda_{i,j-1}\left(A_{j-1} + \bar{g}_{j-1}\hat{A}_{j-1}\right)^T + \left(M_{i,j-1}^T - \sum_{l=0}^{L} \Lambda_{i,j-1-l}\tilde{C}_{j-1-l}^T \bar{F}_{j-1-l}^T\right.$$
$$\left. \times \bar{P}_{i-1-l}^T \right) L_{j-1}^T,$$

$$\Gamma_{i,j} = \left(A_{i-1} + \bar{g}_{i-1}\hat{A}_{i-1}\right)\Gamma_{i-1,j-1}\left(A_{j-1} + \bar{g}_{j-1}\hat{A}_{j-1}\right)^T + L_{i-1}\left(U_{i-1,j-1}\right.$$
$$\left. - \sum_{l=0}^{L} \bar{P}_{l,i-1}\bar{F}_{i-1-l}\tilde{C}_{i-1-l}\Gamma_{i-1-l,j-1}\right)\left(A_{j-1} + \bar{g}_{j-1}\hat{A}_{j-1}\right)^T + \left(A_{i-1}\right.$$
$$\left. + \bar{g}_{i-1}\hat{A}_{i-1}\right)\left(U_{j-1,i-1}^T - \sum_{l=0}^{L} \Gamma_{i-1,j-1-l}\tilde{C}_{j-1-l}^T \bar{F}_{j-1-l}^T \bar{P}_{j-1-l}^T\right)L_{j-1}^T$$
$$+ L_{i-1}\left(R_{i-1,j-1} - \sum_{l=0}^{L} U_{i-1,j-1-l}\tilde{C}_{j-1-l}^T \bar{F}_{j-1-l}^T \bar{P}_{l,j-1}^T - \sum_{l=0}^{L} \bar{P}_{l,i-1}\right.$$
$$\left. \times \bar{F}_{i-1-l}\tilde{C}_{i-1-l}U_{j-1,i-1-l}^T + \sum_{p=0}^{L}\sum_{q=0}^{L} \bar{P}_{p,i-1}\bar{F}_{i-1-p}\tilde{C}_{i-1-p}\Gamma_{i-1-p,j-1-q}\right.$$
$$\left. \times \tilde{C}_{j-1-l}^T \bar{F}_{j-1-l}^T \bar{P}_{l,j-1}^T\right) L_{j-1}^T,$$

$$S_k = \left[M_{k,k}^T - U_{k,k} - \sum_{i=0}^{L} \bar{P}_{i,k}\bar{F}_{k-i}\tilde{C}_{k-i}\left(\Lambda_{k,k-i} - \Gamma_{k,k-i}\right)^T\right]\left(A_k + \bar{g}_k\hat{A}_k\right)^T,$$

$$T_k = \sum_{i=0}^{L}\sum_{j=0}^{L} \bar{P}_{i,k}\bar{F}_{k-i}\tilde{C}_{k-i}\Gamma_{k-i,k-j}\tilde{C}_{k-j}^T \bar{F}_{k-j}^T \bar{P}_{j,k}^T - \sum_{i=0}^{L} \bar{P}_{i,k}\bar{F}_{k-i}\tilde{C}_{k-i}U_{k,k-i}^T$$
$$- \sum_{i=0}^{L} U_{k,k-i}\tilde{C}_{k-j}^T \bar{F}_{k-j}^T \bar{P}_{j,k}^T + R_{k,k}.$$

Moreover, the value of $X_{0,0}$ is known according to the assumptions, and the initial values of some other variables are as follows:

$$U_{i,0} = \sum_{l=0}^{L} \bar{P}_{l,i}\bar{F}_{i-l}\tilde{C}_{i-l} \prod_{n=1}^{i-l}\left(A_{i-l-n} + \bar{g}_{i-l-n}\hat{A}_{i-l-n}\right)\mathbb{E}\{x_0\}\mathbb{E}\left\{x_0^T\right\},$$

$$\Lambda_{i,0} = \prod_{n=1}^{i}\left(A_{i-n} + \bar{g}_{i-n}\hat{A}_{i-n}\right)\mathbb{E}\{x_0\}\mathbb{E}\left\{x_0^T\right\},$$

$$\Gamma_{0,0} = \mathbb{E}\{x_0\}\mathbb{E}\left\{x_0^T\right\}.$$

Proof Denote

$$y_k = \left[\left(y_k^{(1)}\right)^T, \ldots, \left(y_k^{(\mu)}\right)^T\right]^T,$$

$$F_k = \text{diag}\left\{f_k^{(1)} I_{m^{(1)}}, \ldots, f_k^{(\mu)} I_{m^{(\mu)}}\right\},$$

$$v_k = \left[\left(v_k^{(1)}\right)^T, \ldots, \left(v_k^{(\mu)}\right)^T\right]^T,$$

Optimal Filter Design

$$n_k = \left[\left(n_k^{(1)}\right)^T, \ldots, \left(n_k^{(\mu)}\right)^T\right]^T,$$

$$\delta(\tau_k, m) = \text{diag}\left\{\delta\left(\tau_k^{(1)}, m\right) I_{m^{(1)}}, \delta\left(\tau_k^{(2)}, m\right) I_{m^{(2)}}, \ldots, \delta\left(\tau_k^{(\mu)}, m\right) I_{m^{(\mu)}}\right\}.$$

From (2.1)–(2.4), we have:

$$\begin{aligned}
\tilde{x}_{k+1} &= x_{k+1} - \hat{x}_{k+1} \\
&= \left(A_k + \bar{g}_k \hat{A}_k\right) \tilde{x}_k + (g_k - \bar{g}_k) \hat{A}_k x_k + w_k \\
&\quad - L_k \left(r_k - \sum_{j=0}^{L} \bar{P}_{j,k} \bar{F}_{k-j} \tilde{C}_{k-j} \hat{x}_{k-j}\right).
\end{aligned} \quad (2.6)$$

We aim to prove that L_k in (2.5) minimizes the filtering error variance $P_{k+1} = \mathbb{E}\left\{\tilde{x}_{k+1}\tilde{x}_{k+1}^T\right\}$ at each time instant k.

Denote $X_{i,j} = \mathbb{E}\left\{x_i x_j^T\right\}$, $Y_{i,j} = \mathbb{E}\left\{y_i y_j^T\right\}$ and $R_{i,j} = \mathbb{E}\left\{r_i r_j^T\right\}$. According to our assumption, $X_{0,0}$ is known.

When $i = j$, we have

$$\begin{aligned}
X_{i,i} &= \mathbb{E}\Big\{\left[\left(A_{i-1} + g_{i-1}\hat{A}_{i-1}\right) x_{i-1} + w_{i-1}\right] \\
&\quad \times \left[\left(A_{i-1} + g_{i-1}\hat{A}_{i-1}\right) x_{i-1} + w_{i-1}\right]^T\Big\} \\
&= A_{i-1} X_{i-1,i-1} A_{i-1}^T + \bar{g}_{i-1}\left(A_{i-1} X_{i-1,i-1} \hat{A}_{i-1}^T + \hat{A}_{i-1} X_{i-1,i-1} A_{i-1}^T\right) \\
&\quad + \tilde{g}_{i-1}\hat{A}_{i-1} X_{i-1,i-1} \hat{A}_{i-1}^T + W_{i-1}.
\end{aligned}$$

When $i > j$, it is clear that the additive noises $w_j, \ldots, w_{i-1}$ and the multiplicative noises $g_j, \ldots, g_{i-1}$ are independent of x_j. Therefore,

$$\begin{aligned}
X_{i,j} &= \mathbb{E}\left\{\left[\left(A_{i-1} + g_{i-1}\hat{A}_{i-1}\right) x_{i-1} + w_{i-1}\right] x_j^T\right\} \\
&= \left(A_{i-1} + \bar{g}_{i-1}\hat{A}_{i-1}\right) \mathbb{E}\left\{x_{i-1} x_j^T\right\} \\
&= \cdots = \prod_{m=1}^{i-j} \left(A_{i-m} + \bar{g}_{i-m}\hat{A}_{i-m}\right) X_{j,j}.
\end{aligned}$$

When $i < j$, similar with above, it follows that

$$X_{i,j} = X_{i,i} \prod_{m=1}^{j-i} \left(A_{j-m} + \bar{g}_{j-m}\hat{A}_{j-m}\right)^T.$$

Next, because the measurement noise is independent of the state, the following two cases of $Y_{i,j}$ can be dealt with readily:

When $i \neq j$,

$$\begin{aligned}
Y_{i,j} &= \mathbb{E}\left\{\left(F_i \tilde{C}_i x_i + v_i\right)\left(F_j \tilde{C}_j x_j + v_j\right)^T\right\} \\
&= \bar{F}_i \tilde{C}_i X_{i,j} \tilde{C}_j^T \bar{F}_j^T;
\end{aligned}$$

When $i = j$,

$$Y_{i,j} = \mathbb{E}\left\{\left(F_i\tilde{C}_i x_i + v_i\right)\left(F_i\tilde{C}_i x_i + v_i\right)^T\right\}$$
$$= \tilde{F}_i \circ \tilde{C}_i X_{i,i} \tilde{C}_i^T + V_i.$$

R is calculated in the following two cases:
When $i = j$,

$$R_{i,i} = \mathbb{E}\left\{\left(\sum_{m=0}^{L} \delta(\tau_i, m) y_{i-m} + n_i\right)\left(\sum_{m=0}^{L} \delta(\tau_i, m) y_{i-m} + n_i\right)^T\right\}$$
$$= \sum_{m=0}^{L} Q_{m,i} \circ Y_{i-m,i-m} + N_i;$$

When $i \neq j$,

$$R_{i,j} = \sum_{m=0}^{L}\sum_{n=0}^{L} \bar{P}_{m,i} Y_{i-m,j-n} \bar{P}_{n,j}^T.$$

It can also be obtained that

$$H_{i,j} = \mathbb{E}\left\{x_i y_j^T\right\} = \mathbb{E}\left\{x_i(\lambda_j C_j x_j + \lambda_j v_j)^T\right\} = m_j X_{i,j} C_j^T,$$

$$M_{i,j} = \mathbb{E}\left\{x_i r_j^T\right\} = \mathbb{E}\left\{x_i(\sum_{m=0}^{L} I_{\{\tau_j = m\}} y_{j-m} + n_j)^T\right\} = \sum_{m=0}^{L} p_{m,j} H_{i,j-m},$$

$$U_{i,j} = \mathbb{E}\left\{r_i \hat{x}_j^T\right\}$$
$$= U_{i,j-1}\left(A_{j-1} + \bar{g}_{j-1}\hat{A}_{j-1}\right)^T$$
$$+ \left(R_{i,j-1} - \sum_{l=0}^{L} U_{i,j-1-l}\tilde{C}_{j-1-l}^T \bar{F}_{j-1-l}^T \bar{P}_{j-1-l}^T\right) L_{j-1}^T,$$

$$\Lambda_{i,j} = \mathbb{E}\left\{x_i \hat{x}_j^T\right\}$$
$$= \Lambda_{i,j-1}\left(A_{j-1} + \bar{g}_{j-1}\hat{A}_{j-1}\right)^T$$
$$+ \left(M_{i,j-1}^T - \sum_{l=0}^{L} \Lambda_{i,j-1-l}\tilde{C}_{j-1-l}^T \bar{F}_{j-1-l}^T \bar{P}_{i-1-l}^T\right) L_{j-1}^T,$$

$$\Gamma_{i,j} = \mathbb{E}\left\{\hat{x}_i \hat{x}_j^T\right\}$$
$$= \left(A_{i-1} + \bar{g}_{i-1}\hat{A}_{i-1}\right)\Gamma_{i-1,j-1}\left(A_{j-1} + \bar{g}_{j-1}\hat{A}_{j-1}\right)^T$$
$$+ L_{i-1}\left(U_{i-1,j-1} - \sum_{l=0}^{L} \bar{P}_{l,i-1}\bar{F}_{i-1-l}\tilde{C}_{i-1-l}\Gamma_{i-1-l,j-1}\right)$$
$$\times \left(A_{j-1} + \bar{g}_{j-1}\hat{A}_{j-1}\right)^T + \left(A_{i-1} + \bar{g}_{i-1}\hat{A}_{i-1}\right)$$

$$\times \left(U_{j-1,i-1}^T - \sum_{l=0}^{L} \Gamma_{i-1,j-1-l} \tilde{C}_{j-1-l}^T \bar{F}_{j-1-l}^T \bar{P}_{j-1-l}^T \right) L_{j-1}^T$$

$$+ L_{i-1} \Bigg(R_{i-1,j-1} - \sum_{l=0}^{L} U_{i-1,j-1-l} \tilde{C}_{j-1-l}^T \bar{F}_{j-1-l}^T \bar{P}_{l,j-1}^T - \sum_{l=0}^{L} \bar{P}_{l,i-1}$$

$$\times \bar{F}_{i-1-l} \tilde{C}_{i-1-l} U_{j-1,i-1-l}^T + \sum_{p=0}^{L} \sum_{q=0}^{L} \bar{P}_{p,i-1} \bar{F}_{i-1-p} \tilde{C}_{i-1-p} \Gamma_{i-1-p,j-1-q}$$

$$\times \tilde{C}_{j-1-l}^T \bar{F}_{j-1-l}^T \bar{P}_{l,j-1}^T \Bigg) L_{j-1}^T.$$

According to the definitions, the initial values of the variables can be calculated as follows:

$$U_{i,j} = \mathbb{E}\left\{r_i \hat{x}_0^T\right\} = \mathbb{E}\left\{\left(\sum_{l=0}^{L} I_{\{\tau_i = l\}} y_{i-l} + n_i\right) \mathbb{E}\left\{x_0^T\right\}\right\}$$

$$= \sum_{l=0}^{L} p_{l,i} \mathbb{E}\{\lambda_{i-l}\} C_{i-l} \mathbb{E}\{x_{i-l}\} \mathbb{E}\left\{x_0^T\right\}$$

$$= \sum_{l=0}^{L} p_{l,i} m_{i-l} C_{i-l} \left(\prod_{n=1}^{i-l} A_{i-l-n}\right) \mathbb{E}\{x_0\} \mathbb{E}\left\{x_0^T\right\},$$

$$\Lambda_{i,0} = \mathbb{E}\left\{x_i \hat{x}_0^T\right\} = \mathbb{E}\{x_i\} \mathbb{E}\left\{x_0^T\right\} = \left(\prod_{n=1}^{i} A_{i-n}\right) \mathbb{E}\{x_0\} \mathbb{E}\left\{x_0^T\right\},$$

$$\Gamma_{0,0} = \mathbb{E}\left\{\hat{x}_0 \hat{x}_0^T\right\} = \mathbb{E}\{x_0\} \mathbb{E}\left\{x_0^T\right\}.$$

Substituting (2.6) into the definition of P_{k+1}, we obtain

$$P_{k+1} = \mathbb{E}\left\{\tilde{x}_{k+1} \tilde{x}_{k+1}^T\right\}$$

$$= \left(A_k + \bar{g}_k \hat{A}_k\right) P_k \left(A_k + \bar{g}_k \hat{A}_k\right)^T + (\tilde{g}_k - \bar{g}_k^2) \hat{A}_k X_{k,k} \hat{A}_k^T - \Big(A_k$$

$$+ \bar{g}_k \hat{A}_k\Big) \left[M_{k,k} - U_{k,k}^T + \sum_{i=0}^{L} (\Gamma_{k,k-i} - \Lambda_{k,k-i}) \tilde{C}_{k-i}^T \bar{F}_{k-i}^T \bar{P}_{i,k}^T\right] L_k^T - L_k$$

$$\times \left[M_{k,k}^T - U_{k,k} + \sum_{i=0}^{L} \bar{P}_{i,k} \bar{F}_{k-i} \tilde{C}_{k-i} (\Gamma_{k,k-i} - \Lambda_{k,k-i})^T\right] \left(A_k + \bar{g}_k \hat{A}_k\right)^T$$

$$+ L_k \Bigg(R_{k,k} - \sum_{i=0}^{L} \bar{P}_{i,k} \bar{F}_{k-i} \tilde{C}_{k-i} U_{k,k-i}^T - \sum_{i=0}^{L} U_{k,k-i} \tilde{C}_{k-i}^T \bar{F}_{k-i}^T \bar{P}_{i,k}^T$$

$$+ \sum_{i=0}^{L} \sum_{j=0}^{L} \bar{P}_{i,k} \bar{F}_{k-i} \tilde{C}_{k-i} \Gamma_{k-i,k-j} \tilde{C}_{k-j}^T \bar{F}_{k-j}^T \bar{P}_{j,k}^T \Bigg) L_k^T + W_k.$$

Noting the definitions of S_k and T_k, and $T_k = T_k^T$, P_{k+1} can be written as

$$P_{k+1} = \left(A_k + \bar{g}_k \hat{A}_k\right) P_k \left(A_k + \bar{g}_k \hat{A}_k\right)^T + \left(\tilde{g}_k - \bar{g}_k^2\right) \hat{A}_k X_{k,k} \hat{A}_k^T$$
$$- S_k^T L_k^T - L_k S_k + L_k T_k L_k^T + W_k$$
$$= (L_k T_k - S_k^T) T_k^{-1} (L_k T_k - S_k^T)^T + \left(A_k + \bar{g}_k \hat{A}_k\right) P_k \left(A_k + \bar{g}_k \hat{A}_k\right)^T$$
$$- S_k^T T_k^{-1} S_k + \left(\tilde{g}_k - \bar{g}_k^2\right) \hat{A}_k X_{k,k} \hat{A}_k^T + W_k.$$

To this end, it is easily seen that P_{k+1} is minimized when $L_k = S_k^T T_k^{-1}$ as in (2.5), where

$$P_{k+1} = \left(A_k + \bar{g}_k \hat{A}_k\right) P_k \left(A_k + \bar{g}_k \hat{A}_k\right)^T - S_k^T T_k^{-1} S_k$$
$$+ \left(\tilde{g}_k - \bar{g}_k^2\right) \hat{A}_k X_{k,k} \hat{A}_k^T + W_k. \qquad (2.7)$$

The proof of this theorem is now complete.

Based on (2.7), it can be seen that the filtering error variance decreases when the noise variances or the sensor degradation rate variance decrease.

Remark 2.3 When the transmission is perfect and there is neither sensor gain degradation nor multiplicative noise (i.e., for any k and i, Prob $\left\{\tau_k^{(i)} = 0\right\} = 1$, $n_k^{(i)} = 0$, $f_k^{(i)} = 1$, $g_k = 1$), the optimal filter for networked systems with stochastic sensor gain degradations can be specialized to the classical Kalman filter.

2.3 Simulation Example

Consider the following discrete linear time-varying networked system [156] with stochastic sensor gain degradations:

$$A_k = \begin{bmatrix} 1.7240 & -0.7788 \\ 1 & 0.05\sin(k) \end{bmatrix}, \quad C_k^{(1)} = \begin{bmatrix} 0.0286, & 0.0264 \end{bmatrix},$$

$$C_k^{(2)} = \begin{bmatrix} 1, & 0 \end{bmatrix}, \quad \hat{A}_k = I_2, \quad w_k = \begin{bmatrix} 0.1 \\ 0 \end{bmatrix} p_k,$$

where p_k, $v_k^{(1)}$, $v_k^{(2)}$, $n_k^{(1)}$ and $n_k^{(2)}$ are uncorrelated white noises, whose variances are 1, 0.25, 0.25, 1, 1, respectively. Multiplicative noise g_k uniformly distributes over $[-0.1, 0.1]$. The initial value of the state x_0 is uniformly distributed over $[-1, 1]$, and therefore $\hat{x}_0 = [0, \ 0]^T$. The sensor gain degradation rates $f_k^{(1)}$ and $f_k^{(2)}$ uniformly distribute, respectively, over $[0.6, 0.8]$

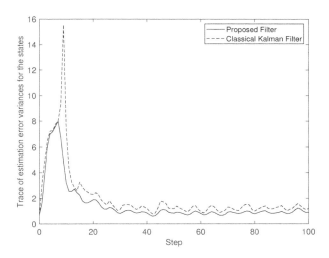

FIGURE 2.1: Trace of estimation error variances for the states

and $[0.5,\ 0.7]$. Furthermore, set the maximum time delay $L=3$, $p_{0,k}^{(i)}=0.4$, $p_{1,k}^{(i)}=0.3$, $p_{2,k}^{(i)}=0.2$, and $p_{3,k}^{(i)}=0.05$ ($i=1,2$) for both transmission channels.

Monte-Carlo simulation is carried out. The commonly used index, trace of estimation error variance, is employed to evaluate the filtering performance. Fig. 2.1 shows the trace of filtering error variance averaged over 150 Monte-Carlos runs of the proposed filter algorithm, and the corresponding simulation result using the traditional Kalman filter method. The simulation ends after 100 steps. It is clear that our proposed filter outperforms the classical Kalman filter, which is due to the efforts we have made in coping with the probabilistic sensor gain degradations and imperfect transmission.

2.4 Conclusions

In this chapter, the optimal filter design problem has been investigated for a class of networked systems with stochastic sensor gain degradations. A linear function of a series of Dirac functions has been introduced to model the imperfect transmission over a shared network with time-varying time delays and data dropouts. μ stochastic sequences $f_k^{(i)}$ have been utilized to describe the random sensor gain degradations. To facilitate online implementation, the parameters of the desired unbiased filter have been calculated recursively to achieve the minimum filtering error variance. Numerical simulation has shown that, in the presence of imperfect transmission and sensor gain degradations, the proposed filter has outperformed the traditional Kalman filter.

3

Recursive Filtering over Sensor Networks with Stochastic Sensor Gain Degradation

Much research effort has recently been devoted to sensor networks due to their successful applications in information collection, environmental monitoring, industrial automation, intelligent buildings, etc. Each sensor node has wireless communication capability and some level of intelligence for signal processing. Via exchanging information with the neighboring nodes, the sensor nodes can jointly fulfill some global tasks. Distributed filtering has been one of the most focused problems in the field of sensor networks, and its main idea is to estimate the dynamics of the target plant based on the distributed nodes. Compared with the traditional single sensor leading to the traditional central filtering approaches [5, 6, 87, 89], each sensor in sensor networks estimates the states of the dynamic process based on not only its own measurement but also the measurements from its neighboring nodes. As such, one of the main difficulties in the distributed filter design problem over a sensor network with given topology is how to take the topology information into account by tackling the complicated couplings between the nodes.

So far, the distributed filtering problem has been gaining an increasing research interest and a wealth of literature has been reported in this topic [11]. For example, the consensus strategy has been applied in Distributed Kalman Filters (DKFs) [80] that allow the nodes in a sensor network to track the average of the sensor measurements based on consensus filters. Communication complexity and packet-loss issues have been discussed for the performance analysis for DKFs in [126]. The optimal distributed filter has been proposed to minimize the filtering error variance [14]. In these optimal filters, the parameters of estimators have been adjusted at each time step to achieve the minimum mean-square estimation error based on the received signals. The filtering algorithm guaranteeing the desired H_∞ performance has been put forward in [187]. Note that, in most of the reported results, the target plants have been limited to the *time-invariant* systems, where the filter performances (e.g. boundedness and monotonicity) have not been investigated in a quantitative way. On the other hand, minimum variance filter has been extensively investigated to achieve the optimal statistical performance. Therefore, there is a practical need to address the distributed filtering problem for *time-varying* systems in sensor networks with detailed analysis on the filter performances including *minimum variance, boundedness, and monotonicity*.

In many practical applications, the phenomenon of sensor gain degradation occurs frequently in a random way. This is particularly true for systems which experience unsteady or abnormal working conditions [112, 148], for example, intermittent sensor outages, sensor aging, or transmission congestions in networked environments. Note that the filtering problem for systems whose sensor gains are subject to random degradation has received some initial research attention [61]. Unfortunately, despite its practical significance, the filtering problem with stochastic sensor gain degradation over sensor networks has not been studied yet for time-varying systems due mainly to the mathematical difficulties, not to mention the case where the filter performance becomes a concern in the design. Those filter structures and design methods in existing literature are not directly applicable to the addressed filtering problem with multiplicative disturbances in the minimum variance sense. *The resulting difficulties stem from the facts that: 1) it is challenging to design an adequate filter gain structure in order to guarantee the minimum error variance at each time step over a sensor network if its topology is not completely connected (i.e. sparse); 2) it is novel to examine how the filter performance is influenced by the statistical law of the sensor gain degradation in a mathematically rigorous way (i.e. monotonicity); 3) it is interesting to establish sufficient conditions under which the estimation error is exponentially bounded in mean square; and 4) it is non-trivial to include the statistical information on the sensor gain degradation in the filter design.* It is, therefore, the main purpose of this chapter to handle the challenges mentioned above by launching a major study on algorithm design and performance analysis issues for recursive filter design problems over possibly sparse sensor networks.

In this chapter, the minimum variance filtering problem is addressed for a class of time-varying systems through sensor networks with stochastic sensor gain degradation. A sequence is employed to describe the gain degradation for each individual sensor. The topology of the sensor network is represented by a directed graph. The minimum-variance distributed filter is designed at each time step using a novel recursive algorithm. The corresponding filtering performances are then analyzed with respect to the boundedness and the monotonicity. Sufficient conditions are obtained under which the estimation error is exponentially bounded in mean square, and the monotonicity property for the error variance with respect to the sensor gain degradation is discussed in the case where all the sensor gains are subject to degradation with the same possibility. Some simulation examples are employed to show the effectiveness of the proposed filtering scheme. *The main contributions of the chapter are outlined as follows: 1) a distributed filter is designed that minimizes the filtering error variance in the presence of stochastic sensor gain degradation; 2) the developed filter caters for time-varying systems in sensor networks, and the algorithm is recursive and thus applicable for online computation; and 3) the estimation performance is investigated, including the analysis of the boundedness and monotonicity of the filtering error dynamics.*

3.1 Problem Formulation and Preliminaries

Consider a sensor network whose topology is represented by a directed graph $\mathcal{G} = (\mathcal{V}, \mathcal{E}, \mathcal{A})$ of order n with the set of nodes $\mathcal{V} = \{1, 2, \ldots, n\}$, set of edges $\mathcal{E} \subseteq \mathcal{V} \times \mathcal{V}$, and a weighted adjacency matrix $\mathcal{A} = [a_{ij}]$ with nonnegative adjacency elements a_{ij}. An edge of $\mathcal{G}$ is denoted by (i, j). The adjacency elements associated with the edges of the graph are positive, i.e., $a_{ij} > 0 \iff (i,j) \in \mathcal{E}$. Moreover, we assume $a_{ii} = 1$ for all $i \in \mathcal{V}$. The set of neighbors of node i plus the node itself are denoted by $\mathcal{N}_i = \{j \in \mathcal{V} : a_{ij} > 0\}$. A communication graph $\mathcal{G}$ is said to be *completely connected* if for any $i, j \in \mathcal{V}$, $(i, j) \in \mathcal{E}$.

Consider the following class of linear discrete time-varying systems

$$x(k+1) = \left[A(k) + \rho(k)\tilde{A}(k)\right] x(k) + w(k), \tag{3.1}$$

where $x(k) \in \mathbb{R}^{n_x}$ is the state; $A(k)$ and $\tilde{A}(k)$ are known matrices with appropriate dimensions; $w(k) \in \mathbb{R}^{n_x}$ is the additive white noise with $\mathbb{E}\{w(k)\} = 0$ and $\mathbb{E}\{w(k)w^T(k)\} = S(k)$. $\rho(k) \in \mathbb{R}$ is the multiplicative noise with $\mathbb{E}\{\rho(k)\} = 0$ and $\mathbb{E}\{\rho^2(k)\} = \xi(k)$. $\mathbb{E}\{x(0)\}$ and $\mathbb{E}\{x(0)x^T(0)\}$ are assumed to be known.

For every sensor node i ($i = 1, 2, \ldots, n$), the measurement is described by

$$y_i(k) = \delta_i(k)C_i(k)x(k) + v_i(k), \tag{3.2}$$

where $y_i(k) \in \mathbb{R}^{n_y}$ is the measurement of the ith node; $C_i(k)$s are known matrices with appropriate dimensions for all $i = 1, 2, \ldots, n$; $v_i(k) \in \mathbb{R}^{n_y}$ is the additive white noise of the ith node with $\mathbb{E}\{v_i(k)\} = 0$ and $\mathbb{E}\{v_i(k)v_i^T(k)\} = V_i(k) > 0$. $\delta_i(k)$, representing the sensor gain degradation in the ith node, is a random variable distributed over the interval $[a_i, b_i]$ ($0 \leq a_i \leq b_i \leq 1$) with $\mathbb{E}\{\delta_i(k)\} = m_i(k)$ and $\text{Var}\{\delta_i(k)\} = l_i(k)$, where $m_i(k)$ and $l_i(k)$ are known scalars. $\rho(k)$, $w(k)$, $\delta_i(k)$ and $v_i(k)$ are all mutually independent.

Let $\hat{x}_i(k) \in \mathbb{R}^{n_x}$ denote the state estimate of the target plant *from the ith node*. In this chapter, the filter to be designed is of the following structure for sensor node i:

$$\hat{x}_i(k+1) = A(k)\hat{x}_i(k) + \sum_{j \in \mathcal{N}_i} H_{ij}(k)a_{ij}\left[y_j(k) - m_j(k)C_j(k)\hat{x}_j(k)\right], \tag{3.3}$$

where the matrices $H_{ij}(k)$ are parameters to be determined. The initial values is $\hat{x}_i(0) = \mathbb{E}\{x(0)\}$ for all $1 \leq i \leq n$. Note that the proposed structure (3.3) reflects how the sensor nodes communicate with their neighbors via $\mathcal{N}_i$ so as to guarantee the unbiased estimation. The unbiasedness can be proven by mathematical induction as follows.

Firstly, one can verify that the unbiasedness assertion is true for $k = 0$ according to $\hat{x}_i(0) = \mathbb{E}\{x(0)\}$ for all $i \in \mathcal{V}$. Secondly, we assume that it is true for the integers from 0 to k and all the $j \in \mathcal{V}$. Letting $\tilde{x}_i(k) = x(k) - \hat{x}_i(k)$, we have the following system that governs the filtering error dynamics:

$$\tilde{x}_i(k+1) = A(k)\tilde{x}_i(k) - \sum_{j \in \mathcal{N}_i} H_{ij}(k)a_{ij}m_j(k)C_j(k)\tilde{x}_j(k)$$

$$- \sum_{j \in \mathcal{N}_i} H_{ij}(k)a_{ij}v_j(k) + \left\{ \rho(k)\tilde{A}(k) \right.$$

$$\left. - \sum_{j \in \mathcal{N}_i} H_{ij}(k)a_{ij}[\lambda_j(k) - m_j(k)]C_j(k) \right\} x(k) + w(k), \qquad (3.4)$$

for $i = 1, 2, \ldots, n$. Then for any $i \in \mathcal{V}$, it follows from (3.4) that

$$\mathbb{E}\{\tilde{x}_i(k+1)\} = A(k)\mathbb{E}\{\tilde{x}_i(k)\} - \sum_{j \in \mathcal{N}_i} H_{ij}(k)a_{ij}m_j(k)C_j(k)\mathbb{E}\{\tilde{x}_j(k)\}$$

$$- \sum_{j \in \mathcal{N}_i} H_{ij}(k)a_{ij}\mathbb{E}\{v_j(k)\} + \left\{ \mathbb{E}\{\rho(k)\}\tilde{A}(k) - \sum_{j \in \mathcal{N}_i} H_{ij}(k)a_{ij} \right.$$

$$\left. \times [\mathbb{E}\{\lambda_j(k)\} - m_j(k)]C_j(k) \right\} \mathbb{E}\{x(k)\} + \mathbb{E}\{w(k)\}.$$

Then, considering the facts that $w(k)$, $\rho(k)$, $\tilde{x}_i(k)$, and $v_i(k)$ are all zero-mean, and $\mathbb{E}\{\lambda_j(k)\} = m_j(k)$, it can be concluded that $\mathbb{E}\{\tilde{x}_i(k+1)\} = 0$ for any $i \in \mathcal{V}$. That concludes the proof. For notational simplicity, we define

$$\bar{x}(k) = \text{vec}_n^T\{\tilde{x}_i^T(k)\}, \quad \bar{x}(k) = \text{vec}_n^T\{x^T(k)\},$$
$$\bar{M}(k) = \text{diag}_n\{m_i(k)I\}, \quad H(k) = [H_{ij}(k)]_{n \times n},$$
$$\bar{C}(k) = \text{diag}_n\{C_i(k)\}, \quad \bar{A}(k) = \text{diag}_n\{A(k)\},$$
$$\bar{w}(k) = \text{vec}_n^T\{w^T(k)\}, \quad \bar{v}(k) = \text{vec}_n^T\{v_i^T(k)\},$$
$$\hat{A}(k) = \text{diag}_n\{\tilde{A}(k)\}, \quad T_i = \text{diag}\{a_{i1}I, \ldots, a_{in}I\},$$
$$\bar{\Lambda}(k) = \text{diag}_n\{\delta_i(k)I\}, \quad E_i = \text{diag}\{\underbrace{0, \ldots, 0}_{i-1}, I, \underbrace{0, \ldots, 0}_{n-i}\},$$

and then (3.4) can be rewritten in the following form:

$$\tilde{x}(k+1) = \left[\bar{A}(k) - \sum_{i=1}^{n} E_i H(k) T_i \bar{M}(k)\bar{C}(k)\right]\tilde{x}(k) + \left\{\rho(k)\hat{A}(k)\right.$$

$$\left. - \sum_{i=1}^{n} E_i H(k) T_i [\bar{\Lambda}(k) - \bar{M}(k)]\bar{C}(k)\right\}\bar{x}(k)$$

$$- \sum_{i=1}^{n} E_i H(k) T_i \bar{v}(k) + \bar{w}(k). \qquad (3.5)$$

Main Results 43

Defining the error covariance at the kth time step as

$$P(k) = \mathbb{E}\left\{\tilde{x}(k)\tilde{x}^T(k)\right\}, \tag{3.6}$$

the goal of this chapter can be stated as designing a filter of the form (3.3) for system (3.1)–(3.2) so that the filtering error covariance $P(k+1)$ is minimized at each time-step k.

Remark 3.1 *In the proposed filter, each sensor node estimates the system states based on the local measurement and measurements of its neighboring sensors. The structure of the filter is set so as to achieve the unbiased estimation at each node. The statistics exploited in the filter (i.e., $m_i(k)$ for all $i \in \mathcal{V}$) are scalars known a priori, which facilitates the filter implementation. In (3.3), the term $y_j(k) - m_j(k)C_j(k)\hat{x}_j(k)$ can represent the difference between the estimated output and actual output of the jth node. In the presence of the complicated interconnections between the sensor nodes, there are various filter structures that can be adopted to achieve unbiased distributed state estimation and the proposed structure (3.3) is among them.*

3.2 Main Results

3.2.1 Filter Design

In this subsection, a set of recursive Riccati-like matrix equations are derived to calculate the filter parameters $H_{ij}(k)$ in (3.3) in order to minimize the error variance for system (3.1). The following theorem gives the parameterizations of the desired filter gains in the two cases that the network is completely or not completely connected.

For presentation convenience, we denote

$$\Omega(0) := \mathbb{E}\left\{x(0)x^T(0)\right\}, \tag{3.7}$$

$$\Omega(k) := \mathbb{E}\left\{x(k)x^T(k)\right\}, \tag{3.8}$$

$$\bar{\Omega}(k) := \mathbb{E}\left\{\bar{x}(k)\bar{x}^T(k)\right\} = [\Omega(k)]_{n \times n}, \tag{3.9}$$

$$U(k) := \mathbb{E}\left\{\left[\bar{\Lambda}(k) - \bar{M}(k)\right]\bar{C}(k)\bar{x}(k)\bar{x}^T(k)\bar{C}^T(k)\left[\bar{\Lambda}(k) - \bar{M}(k)\right]^T\right\}$$

$$= \text{diag}_n\left\{l_i(k)C_i(k)\Omega(k)C_i^T(k)\right\}, \tag{3.10}$$

$$W(k) := \mathbb{E}\left\{\bar{w}(k)\bar{w}^T(k)\right\} = [S(k)]_{n \times n}, \tag{3.11}$$

$$V(k) := \mathbb{E}\left\{\bar{v}(k)\bar{v}^T(k)\right\} = \text{diag}_n\{V_i(k)\}, \tag{3.12}$$

$$Y(k) := \bar{M}(k)\bar{C}(k)P(k)\bar{C}^T(k)\bar{M}^T(k) + V(k) + U(k), \tag{3.13}$$

$$Z(k) := \bar{M}(k)\bar{C}(k)P(k)\bar{A}^T(k), \tag{3.14}$$

$$\mathcal{H}(k) := Z^T(k)Y^{-1}(k) = [\mathcal{H}_{ij}(k)]_{n \times n}. \tag{3.15}$$

Theorem 3.1 *The following statements are true:*

a). If the sensor network topology is completely connected, then the parameters of filter (3.3) achieving the minimum filtering error variance are given by:

$$H_{ij}(k) = \mathcal{H}_{ij}(k) a_{ij}^{-1}, \qquad (3.16)$$

and $P(k)$ is calculated as:

$$\begin{aligned}P(k+1) = &- Z^T(k) Y^{-1}(k) Z(k) + \bar{A}(k) P(k) \bar{A}^T(k) \\ &+ \xi(k) \hat{A}(k) \bar{\Omega}(k) \hat{A}^T(k) + W(k).\end{aligned} \qquad (3.17)$$

b). If the sensor network topology is not completely connected, a practical solution for the parameters of filter (3.3) is given by:

$$H_{ij}(k) = \begin{cases} \mathcal{H}_{ij}(k) a_{ij}^{-1}, & \text{if } a_{ij} \neq 0, \\ 0, & \text{if } a_{ij} = 0, \end{cases} \qquad (3.18)$$

and $P(k)$ obeys the following recursion:

$$\begin{aligned}P(k+1) = &\left[\sum_{i=1}^{n} E_i \mathcal{H}(k) T_i - Z^T(k) Y^{-1}(k) \right] Y(k) \\ &\times \left[\sum_{i=1}^{n} E_i \mathcal{H}(k) T_i - Z^T(k) Y^{-1}(k) \right]^T \\ &- Z^T(k) Y^{-1}(k) Z(k) + \bar{A}(k) P(k) \bar{A}^T(k) \\ &+ \xi(k) \hat{A}(k) \bar{\Omega}(k) \hat{A}^T(k) + W(k).\end{aligned} \qquad (3.19)$$

Proof *a). It follows from the definition (3.6) and the notations (3.7)-(3.15) that*

$$\begin{aligned}P(k+1) = &\left[\bar{A}(k) - \sum_{i=1}^{n} E_i H(k) T_i \bar{M}(k) \bar{C}(k) \right] P(k) \\ &\times \left[\bar{A}(k) - \sum_{i=1}^{n} E_i H(k) T_i \bar{M}(k) \bar{C}(k) \right]^T \\ &+ \left[\sum_{i=1}^{n} E_i H(k) T_i \right] V(k) \left[\sum_{i=1}^{n} E_i H(k) T_i \right]^T \\ &+ \left[\sum_{i=1}^{n} E_i H(k) T_i \right] U(k) \left[\sum_{i=1}^{n} E_i H(k) T_i \right]^T \\ &+ \xi(k) \hat{A}(k) \bar{\Omega}(k) \hat{A}^T(k) + W(k).\end{aligned} \qquad (3.20)$$

Main Results

Moreover, $\Omega(k)$ can be recursively calculated as follows:

$$\begin{aligned}
\Omega(k+1) &= \mathbb{E}\left\{x(k+1)x^T(k+1)\right\} \\
&= \mathbb{E}\bigg\{\left\{\left[A(k) + \rho(k)\tilde{A}(k)\right]x(k) + w(k)\right\} \\
&\quad \times \left\{\left[A(k) + \rho(k)\tilde{A}(k)\right]x(k) + w(k)\right\}^T\bigg\} \\
&= A(k)\Omega(k)A^T(k) + \xi(k)\tilde{A}(k)\Omega(k)\tilde{A}^T(k) + S(k).
\end{aligned} \quad (3.21)$$

From (3.20), it follows that

$$\begin{aligned}
P(k+1) &= \left[\sum_{i=1}^n E_i H(k) T_i\right] Y(k) \left[\sum_{i=1}^n E_i H(k) T_i\right]^T \\
&\quad - Z^T(k)\left[\sum_{i=1}^n E_i H(k) T_i\right]^T - \left[\sum_{i=1}^n E_i H(k) T_i\right] Z(k) \\
&\quad + \xi(k)\hat{A}(k)\bar{\Omega}(k)\hat{A}^T(k) + \bar{A}(k)P(k)\bar{A}^T(k) + W(k).
\end{aligned} \quad (3.22)$$

Since $Y(k) = Y^T(k) > 0$, (3.22) can be rewritten as

$$\begin{aligned}
P(k+1) &= \left[\sum_{i=1}^n E_i H(k) T_i - Z^T(k) Y^{-1}(k)\right] Y(k) \\
&\quad \times \left[\sum_{i=1}^n E_i H(k) T_i - Z^T(k) Y^{-1}(k)\right]^T \\
&\quad - Z^T(k) Y^{-1}(k) Z(k) + \bar{A}(k) P(k) \bar{A}^T(k) \\
&\quad + \xi(k)\hat{A}(k)\bar{\Omega}(k)\hat{A}^T(k) + W(k).
\end{aligned} \quad (3.23)$$

In view of (3.15), it is obvious that $P(k+1)$ is minimized if and only if

$$\sum_{i=1}^n E_i H(k) T_i = \mathcal{H}(k). \quad (3.24)$$

To this end, it can be easily seen that, if the network topology is completely connected, then the minimum variance of the filtering error is achieved when $H(k)$ is calculated as in (3.16), which guarantees

$$[H_{i1}(k), \ldots, H_{in}(k)] = [\mathcal{H}_{i1}(k), \ldots, \mathcal{H}_{in}(k)] T_i^{-1}.$$

Furthermore, base on (3.6), the initial value of P is given by

$$\begin{aligned}
P(0) &= \mathbb{E}\left\{\tilde{x}(0)\tilde{x}^T(0)\right\} \\
&= \left[\Omega(0) - \mathbb{E}\left\{x(0)\right\}\mathbb{E}\left\{x^T(0)\right\}\right]_{n \times n},
\end{aligned}$$

and then $P(k)$ can be updated according to (3.17).

b). In the case that the sensor network topology is not completely connected, it is easily seen that T_i is not invertible for all the $i \in \mathcal{V}$, and therefore the condition (3.24) no longer holds because of the sparse topology. For such a circumstance, an alternative yet effective way for designing the filter gains is to calculate $H(k)$ as (3.18). In doing so, it can be guaranteed that

$$[H_{i1}(k),\ldots,H_{in}(k)] = [\mathcal{H}_{i1}(k),\ldots,\mathcal{H}_{in}(k)]\, T_i^\dagger, \qquad (3.25)$$

where $T_i^\dagger$ is the Moore-Penrose pseudo inverse of T_i. Moreover, in this case, it is straightforward to see that $P(k)$ can be recursively determined as (3.19). This ends the proof.

Remark 3.2 The filter parameters $H_{ij}(k)$ are calculated at each time step to minimize the filtering error covariance. It is worth mentioning that some statistics of the stochastic sensor gain degradations and noises, the network topology, the state transition matrix, and the measurement matrices are required to determine $P(k)$ and $H_{ij}(k)$. To update $P(k)$ and $H_{ij}(k)$ at each node, it is not necessary to request global measurements from all the sensors. Thus, $P(k)$ and $H_{ij}(k)$ can be updated at each node and the proposed algorithm is applicable in distributed sensor networks. It should also be noted that the filter designed in Part b) of Theorem 3.1 is obtained without assuming the complete connectedness of the network topology. In fact, the distance between the proposed filter parameter and the optimal solution $\mathcal{H}(k)$ is minimized in the sense of Euclidean norm, and the proposed filter is applicable in a sparse sensor network.

Remark 3.3 The inverse of $Y(k)$ is utilized to design the desired filter. From (3.13), it is easy to see that $Y(k) = Y^T(k) \geq 0$. Only when all the additive and multiplicative noises are zero, and the missing measurements probability is exactly known, $Y(k)$ may be positive semidefinite, otherwise $Y(k)$ is positive definite. Since noises are unavoidable in most practical cases, $Y(k)$ can be regarded invertible.

In the next subsection, we proceed to deal with the performance evaluation problem of the filter developed in Theorem 3.1. Let us first discuss the boundedness of the estimation error at each time step.

3.2.2 Boundedness

For the dynamics analysis of the estimation error, we will need the following two widely used concepts for the boundedness of stochastic processes [2,158].

Definition 3.1 The stochastic process $\zeta(k)$ is said to be exponentially bounded in mean square if there are real numbers $\eta > 0$, $\nu > 0$ and $0 < \vartheta < 1$ such that

$$\mathbb{E}\left\{\|\zeta(k)\|^2\right\} \leq \eta \|\zeta(0)\|^2 \vartheta^k + \nu \qquad (3.26)$$

holds for every $k > 0$.

Main Results

Definition 3.2 *The stochastic process $\zeta(k)$ is said to be bounded with probability one if*
$$\sup_{k\geq 0}\|\zeta(k)\| < \infty \qquad (3.27)$$
is true with probability one.

For the boundedness of the estimation error, we will first establish sufficient conditions under which the estimation error is exponentially bounded in mean square, and then generalize the results to the case of boundedness with probability one. For this purpose, as in [136], we make the following two assumptions.

Assumption 3.1 *There are positive real numbers $\bar{a}_1, \bar{a}_2, \bar{c}, \underline{c}, \bar{\omega}, \bar{\xi}, \bar{l}, \bar{m}, \underline{m}, \bar{w}, \underline{w}, \underline{v}, \bar{v} > 0$ such that the following bounds on various matrices are fulfilled for every $1 \leq i \leq n$ and $k \geq 0$:*

$$\|A(k)\| \leq \bar{a}_1, \quad \left\|\hat{A}(k)\right\| \leq \bar{a}_2, \quad \underline{c} \leq \|C_i(k)\| \leq \bar{c}, \quad \operatorname{tr}\{\bar{\Omega}(k)\} \leq \bar{\omega}, \quad \xi(k) \leq \bar{\xi},$$
$$\underline{m}I \leq \bar{M}(k) \leq \bar{m}I, \quad l_i(k) \leq \bar{l}, \quad \underline{w}I \leq W(k) \leq \bar{w}I, \quad \underline{v}I \leq V(k) \leq \bar{v}I. \qquad (3.28)$$

Moreover, the following inequality holds:
$$\bar{a}_1 \left(1 + \frac{n\bar{m}^2\bar{c}^2}{\underline{m}^2\underline{c}^2}\right) < 1. \qquad (3.29)$$

Assumption 3.2 *There exists a positive real number $\epsilon > 0$ such that the initial estimation error satisfies*
$$\|\tilde{x}(0)\| \leq \epsilon. \qquad (3.30)$$

Theorem 3.2 *Consider the time-varying system (3.1)–(3.2) with the minimum-variance filter given in (3.3) whose parameters are provided in Theorem 3.1. Under Assumption 3.1, the estimation error given by (3.4) is exponentially bounded in mean square.*

Proof *Denoting*
$$\check{A}(k) = \bar{A}(k) - \sum_{i=1}^{n} E_i H(k) T_i \bar{M}(k) \bar{C}(k),$$
then (3.5) can be written as
$$\tilde{x}(k+1) = \check{A}(k)\tilde{x}(k) + r(k) + s(k), \qquad (3.31)$$
where
$$r(k) = \left\{\rho(k)\hat{A}(k) - \sum_{i=1}^{n} E_i H(k) T_i \left[\bar{\Lambda}(k) - \bar{M}(k)\right]\bar{C}(k)\right\}\bar{x}(k),$$
$$s(k) = -\sum_{i=1}^{n} E_i H(k) T_i \bar{v}(k) + \bar{w}(k). \qquad (3.32)$$

Based on (3.15), it follows easily from $\|\bar{A}(k)\| \leq \bar{a}_1$ and $\underline{c} \leq \|\bar{C}(k)\| \leq \bar{c}$ that

$$\|\mathcal{H}(k)\| = \left\|\bar{A}(k)P(k)\bar{C}^T(k)\bar{M}^T(k)[\bar{M}(k)\bar{C}(k)P(k)\bar{C}^T(k)\bar{M}^T(k) \right.$$
$$\left. + V(k) + U(k)]^{-1}\right\|$$
$$< \frac{\bar{a}_1 \bar{m} \bar{c}}{\underline{m}^2 \underline{c}^2} := \bar{h}. \tag{3.33}$$

According to the fact that $\left\|\sum_{i=1}^n E_i H(k) T_i\right\| \leq n\|\mathcal{H}(k)\|$, we have

$$\|\check{A}(k)\| \leq \|\bar{A}(k)\| + \left\|\sum_{i=1}^n E_i H(k) T_i\right\| \|\bar{M}(k)\bar{C}(k)\|$$
$$\leq \|\bar{A}(k)\| + \|\mathcal{H}(k)\| \|\bar{M}(k)\bar{C}(k)\|$$
$$< \bar{a}_1 + n\bar{h}\bar{m}\bar{c} := \bar{a}. \tag{3.34}$$

Based on (3.29) and (3.33), it can be seen that $\bar{a} < 1$.

Denoting $L(k) = \text{diag}_n\{l_i(k)I\}$, it is obvious that $L(k) < \bar{l}I$. Furthermore, we have

$$\left\|\mathbb{E}\left\{\left\{\rho(k)\hat{A}(k) - \sum_{i=1}^n E_i H(k) T_i[\bar{\Lambda}(k) - \bar{M}(k)]\bar{C}(k)\right\}\right.\right.$$
$$\left.\left.\times \left\{\rho(k)\hat{A}(k) - \sum_{i=1}^n E_i H(k) T_i[\bar{\Lambda}(k) - \bar{M}(k)]\bar{C}(k)\right\}^T\right\}\right\|$$
$$= \left\|\xi(k)\hat{A}(k)\hat{A}^T(k) + \left[\sum_{i=1}^n E_i H(k) T_i\right][L(k)\bar{C}(k)\bar{C}^T(k)]\left[\sum_{i=1}^n E_i H(k) T_i\right]^T\right\|$$
$$\leq \left\|\xi(k)\hat{A}(k)\hat{A}^T(k)\right\| + n^2\|\mathcal{H}(k)\|^2 \|L(k)\bar{C}(k)\bar{C}^T(k)\|$$
$$< \bar{\xi}\bar{a}_2^2 + n^2\bar{h}^2\bar{l}\bar{c}^2. \tag{3.35}$$

Thus, it is clear that

$$\mathbb{E}\left\{r^T(k)r(k)\right\} < \left(\bar{\xi}\bar{a}_2^2 + n^2\bar{h}^2\bar{l}\bar{c}^2\right)\bar{\omega}^2 := \bar{r}^2, \tag{3.36}$$

and the bounds of $s(k)$ can be calculated as follows:

$$\mathbb{E}\left\{s^T(k)s(k)\right\}$$
$$= \mathbb{E}\left\{\bar{v}^T(k)\left[\sum_{i=1}^n E_i H(k) T_i\right]^T \left[\sum_{i=1}^n E_i H(k) T_i\right]\bar{v}(k) + \bar{w}^T(k)\bar{w}(k)\right\}$$
$$< \mathbb{E}\left\{n^2\bar{h}^2\bar{v}^T(k)\bar{v}(k) + \bar{w}^T(k)\bar{w}(k)\right\}$$
$$= \mathbb{E}\left\{\text{tr}\left\{n^2\bar{h}^2\bar{v}(k)\bar{v}^T(k) + \bar{w}(k)\bar{w}^T(k)\right\}\right\}$$
$$\leq n^3 n_y \bar{h}^2 \bar{v} + n n_x \bar{w} := \bar{s}^2. \tag{3.37}$$

Main Results 49

$$\mathbb{E}\left\{s(k)s^T(k)\right\} \geq \mathbb{E}\left\{\bar{w}(k)\bar{w}^T(k)\right\} = W(k) \geq \underline{w}I := \underline{s}^2 I. \tag{3.38}$$

From (3.31), it follows that

$$P(k+1) = \check{A}(k)P(k)\check{A}^T(k) + \check{R}(k) + \check{S}(k), \tag{3.39}$$

where

$$\check{R}(k) = \mathbb{E}\left\{r(k)r^T(k)\right\}, \quad \check{S}(k) = \mathbb{E}\left\{s(k)s^T(k)\right\}.$$

It is straightforward to see that $\check{R}(k) \leq \bar{r}^2 I$ and $\underline{s}^2 I \leq \check{S}(k) \leq \bar{s}^2 I$.

Consider the following iterative matrix equation with respect to $\Pi(k)$:

$$\begin{aligned}\Pi(k+1) =& \check{A}(k)\Pi(k)\check{A}^T(k) + \left[\sum_{i=1}^{n} E_i H(k)T_i\right]V(k) \\ & \times \left[\sum_{i=1}^{n} E_i H(k)T_i\right]^T + W(k).\end{aligned} \tag{3.40}$$

with initial condition

$$\Pi(0) = \left[\sum_{i=1}^{n} E_i H(0)T_i\right]V(0)\left[\sum_{i=1}^{n} E_i H(0)T_i\right]^T + W(0).$$

With the definition of $\check{S}(k)$, (3.40) can be rewritten as:

$$\Pi(k+1) = \check{A}(k)\Pi(k)\check{A}^T(k) + \check{S}(k). \tag{3.41}$$

Then, it follows directly that

$$\|\Pi(k+1)\| \leq \|\Pi(k)\|\left\|\check{A}(k)\right\|^2 + \left\|\check{S}(k)\right\| \leq \bar{a}^2 \|\Pi(k)\| + \bar{s}^2. \tag{3.42}$$

By iteration, we have

$$\|\Pi(k)\| \leq \bar{a}^{2k}\|\Pi(0)\| + \bar{s}^2 \sum_{i=0}^{k-1} \bar{a}^{2i}. \tag{3.43}$$

Since $0 < \bar{a} < 1$, we arrive at

$$\|\Pi(k)\| < \|\Pi(0)\| + \bar{s}^2 \sum_{i=0}^{\infty} \bar{a}^{2i} = \|\Pi(0)\| + \frac{\bar{s}^2}{1-\bar{a}^2}. \tag{3.44}$$

Furthermore, since $\Pi(k)$ is positive definite for all k, it is straightforward to see that

$$\|\Pi(k+1)\| \geq \left\|\check{S}(k)\right\| \geq \underline{s}^2. \tag{3.45}$$

Based on (3.44) and (3.45), it can be concluded that there are positive real numbers $\underline{\pi}, \bar{\pi} > 0$ such that the inequality $\underline{\pi} I \leq \Pi(k) \leq \bar{\pi} I$ holds for every $k \geq 0$.

According to (3.41), we obtain

$$\check{A}^T(k)\Pi^{-1}(k+1)\check{A}(k) - \Pi^{-1}(k)$$
$$= -\left\{\Pi(k) + \Pi(k)\check{A}^T(k)\check{S}^{-1}(k)\check{A}(k)\Pi(k)\right\}^{-1}$$
$$= -\left\{\check{A}^T(k)\check{S}^{-1}(k)\check{A}(k)\Pi(k) + I\right\}^{-1}\Pi^{-1}(k)$$
$$< -\left(\frac{\bar{a}^2 \bar{\pi}}{\underline{s}^2} + 1\right)^{-1}\Pi^{-1}(k). \tag{3.46}$$

Defining $\alpha = \left(\bar{a}^2\bar{\pi}/\underline{s}^2 + 1\right)^{-1}$, $\mu = \left(\bar{r}^2 + \bar{s}^2\right)/\underline{\pi}$, and $V_k(\tilde{x}(k)) = \tilde{x}(k)^T \Pi^{-1}(k)\tilde{x}(k)$, we obtain from (3.31) that

$$\mathbb{E}\left\{V_{k+1}(\tilde{x}(k+1))|\tilde{x}(k)\right\} - V_k(\tilde{x}(k))$$
$$= \mathbb{E}\left\{\left[\check{A}(k)\tilde{x}(k) + r(k) + s(k)\right]^T \Pi^{-1}(k+1)\right.$$
$$\left. \times \left[\check{A}(k)\tilde{x}(k) + r(k) + s(k)\right]\right\} - \tilde{x}^T(k)\Pi^{-1}(k)\tilde{x}(k)$$
$$= -\tilde{x}^T(k)\left\{\check{A}^T(k)\check{S}^{-1}(k)\check{A}(k)\Pi(k) + I\right\}^{-1}\Pi^{-1}(k)\tilde{x}(k)$$
$$+ \mathbb{E}\left\{r^T(k)\Pi^{-1}(k+1)r(k)\right\} + \mathbb{E}\left\{s^T(k)\Pi^{-1}(k+1)s(k)\right\}$$
$$< -\alpha V_k(\tilde{x}(k)) + \mu, \tag{3.47}$$

which gives rise to

$$\mathbb{E}\left\{\|\tilde{x}(k)\|^2\right\} \leq \frac{\bar{\pi}}{\underline{\pi}} \|\tilde{x}(0)\|^2 (1-\alpha)^k + \mu\bar{\pi} \sum_{i=0}^{k-1}(1-\alpha)^i. \tag{3.48}$$

Noticing $0 < \alpha < 1$ and $\mu, \bar{\pi} > 0$, it follows that

$$\mathbb{E}\left\{\|\tilde{x}(k)\|^2\right\} \leq \frac{\bar{\pi}}{\underline{\pi}} \|\tilde{x}(0)\|^2 (1-\alpha)^k + \mu\bar{\pi} \sum_{i=0}^{\infty}(1-\alpha)^i$$
$$= \frac{\bar{\pi}}{\underline{\pi}} \|\tilde{x}(0)\|^2 (1-\alpha)^k + \frac{\mu\bar{\pi}}{\alpha}. \tag{3.49}$$

Therefore, the stochastic process $\tilde{x}(k)$ is exponentially bounded in mean square and the proof is complete.

Under Assumption 3.2, i.e., the initial filtering error is bounded, we have $\|\tilde{x}(0)\|^2 < \epsilon < \infty$. Then, it follows from (3.30) and (3.49) that the stochastic process $\tilde{x}(k)$ is bounded with probability 1. In this case, the following corollary is easily accessible.

Main Results 51

Corollary 3.1 *Consider the time-varying system (3.1)–(3.2) with the minimum-variance filter given in (3.3) whose parameters are provided in Theorem 3.1. Under Assumptions 3.1–3.2, the estimation error given by (3.4) is bounded with probability 1.*

Remark 3.4 *The main proof of Theorem 3.2 provides a constructive way to quantify the error bound in (3.26). Furthermore, in practical systems, because of the energy constraints, it is reasonable to assume that the noise variances and the spectral norms of transfer matrix and measurement matrix are bounded.*

Remark 3.5 *Compared with the results in [136], the bound obtained in Theorem 3.2 is obviously dependent on the stochastic sensor gain degradation and the network topology, and this reflects our efforts on dealing with the stochastic and distributed nature of the system. Therefore, Theorem 3.2 offers a more applicable error bound to systems over sensor networks with stochastic sensor gain degradations. The consideration of the distributed measurements and the verifiable condition (3.29) constitute the main differences between our boundedness analysis and those in [83, 136]. It is worth mentioning that to reduce the conservativeness of the sufficient conditions that can guarantee the boundedness of the estimation error would be one of the future research directions.*

3.2.3 Monotonicity

In this part, we aim to discuss the relationship between the filtering performance and the sensor gain degradation. For demonstration purpose, we assume that all $\delta_i(k)$ have the same statistics, i.e., $\bar{M}(k) = m(k)I$. We utilize $\operatorname{tr}\{P(k)\}$ as a standard criterion to measure the filtering performance.

In the following theorem, the influence from $m(k)$ to $\operatorname{tr}\{P(k+1)\}$ is clearly revealed.

Theorem 3.3 $\operatorname{tr}\{P(k+1)\}$ *is nonincreasing when $m(k)$ increases.*

Proof *Denote $F_i = \operatorname{diag}\{f_{i1}I, \ldots, f_{in}I\}$ where*

$$f_{ij} = \begin{cases} 0, & \text{if } a_{ij} \neq 0, \\ 1, & \text{if } a_{ij} = 0. \end{cases}$$

Based on our assumption, we have $a_{ii} = 1$ and therefore $F_i < I$ for all $1 \leq i \leq n$.

From the definitions of E_i, T_i, F_i and Theorem 3.1, it is clear that

$$\sum_{i=1}^{n} E_i H(k) T_i - \mathcal{H}(k) = -\sum_{i=1}^{n} E_i \mathcal{H}(k) F_i, \qquad (3.50)$$

and then it follows from (3.23) and (3.50) that

$$P(k+1) = \left[\sum_{i=1}^{n} E_i \mathcal{H}(k) F_i\right] Y(k) \left[\sum_{j=1}^{n} E_j \mathcal{H}(k) F_j\right]^T$$

$$- \left[\sum_{i=1}^{n} E_i\right] Z^T(k) Y^{-1}(k) Z(k) \left[\sum_{j=1}^{n} E_j\right]^T$$

$$+ \bar{A}(k) P(k) \bar{A}^T(k) + \xi(k) \hat{A}(k) \bar{\Omega}(k) \hat{A}^T(k) + W(k)$$

$$= \sum_{i=1}^{n} \sum_{j=1}^{n} E_i \mathcal{H}(k) F_i Y(k) F_j^T \mathcal{H}^T(k) E_j^T$$

$$- \sum_{i=1}^{n} \sum_{j=1}^{n} E_i Z^T(k) Y^{-1}(k) Z(k) E_j^T$$

$$+ \bar{A}(k) P(k) \bar{A}^T(k) + \xi(k) \hat{A}(k) \bar{\Omega}(k) \hat{A}^T(k) + W(k). \quad (3.51)$$

To simplify (3.51), we notice that, when $i \neq j$, the following facts are true: $\mathrm{tr}\left\{E_i \mathcal{H}(k) F_i Y(k) F_j^T \mathcal{H}^T(k) E_j^T\right\} = 0$, $\mathrm{tr}\left\{E_i Z^T(k) Y^{-1}(k) Z(k) E_j^T\right\} = 0$. Then, we have

$$\mathrm{tr}\left\{P(k+1)\right\} = \mathrm{tr}\Bigg\{\sum_{i=1}^{n} E_i \mathcal{H}(k) F_i Y(k) F_i^T \mathcal{H}^T(k) - \sum_{i=1}^{n} E_i Z^T(k) Y^{-1}(k) Z(k)$$

$$+ \bar{A}(k) P(k) \bar{A}^T(k) + \xi(k) \hat{A}(k) \bar{\Omega}(k) \hat{A}^T(k) + W(k)\Bigg\}. \quad (3.52)$$

Furthermore, it follows from (3.52) that

$$\frac{\partial \mathrm{tr}\left\{P(k+1)\right\}}{\partial m(k)} = \frac{\partial}{\partial m(k)} \mathrm{tr}\Bigg\{\sum_{i=1}^{n} E_i \mathcal{H}(k) F_i Y(k) F_i^T \mathcal{H}^T(k)$$

$$- \sum_{i=1}^{n} E_i Z^T(k) Y^{-1}(k) Z(k)\Bigg\}$$

$$= \frac{\partial}{\partial m(k)} \sum_{i=1}^{n} \mathrm{tr}\Big\{E_i Z^T(k) Y^{-1}(k)$$

$$\times \left[F_i Y(k) F_i^T - Y(k)\right] Y^{-1}(k) Z(k)\Big\}. \quad (3.53)$$

With (3.13), (3.14), and (3.53), we have

$$\frac{\partial \mathrm{tr}\left\{P(k+1)\right\}}{\partial m(k)} = \sum_{i=1}^{n} \mathrm{tr}\Big\{2\bar{C}(k) P(k) \bar{A}^T(k) E_i Z^T(k) Y^{-1}(k) [F_i Y(k) F_i^T$$

Main Results

$$\begin{aligned}
&- Y(k)]Y^{-1}(k) - 2\bar{C}(k)P(k)\bar{C}^T(k)\bar{M}^T(k)Y^{-1}(k)\Big\{[F_iY(k)\\
&\times F_i^T - Y(k)]Y^{-1}(k)Z(k)E_iZ^T(k) + Z(k)E_iZ^T(k)Y^{-1}(k)\\
&\times [F_iY(k)F_i^T - Y(k)]\Big\}Y^{-1}(k) + 2\bar{C}(k)P(k)\bar{C}^T(k)\bar{M}^T(k)\\
&\times \Big[F_i^T Y^{-1}(k)Z(k)E_iZ^T(k)Y^{-1}(k)F_i - Y^{-1}(k)Z(k)E_i\\
&\times Z^T(k)Y^{-1}(k)\Big]\Big\}.
\end{aligned} \tag{3.54}$$

When $m(k) > 0$, we obtain

$$\begin{aligned}
\frac{\partial \mathrm{tr}\{P(k+1)\}}{\partial m(k)} &= \sum_{i=1}^n \frac{1}{m(k)} \mathrm{tr}\Big\{2Z(k)E_iZ^T(k)Y^{-1}(k)\left[F_iY(k)F_i^T - Y(k)\right]\\
&\times Y^{-1}(k) - 2\bar{M}(k)\bar{C}(k)P(k)\bar{C}^T(k)\bar{M}^T(k)Y^{-1}(k)\Big\{F_iY(k)\\
&\times F_i^T Y^{-1}(k)Z(k)E_iZ^T(k) + Z(k)E_iZ^T(k)Y^{-1}(k)F_iY(k)\\
&\times F_i^T\Big\}Y^{-1}(k) + 2\bar{M}(k)\bar{C}(k)P(k)\bar{C}^T(k)\bar{M}^T(k)F_i^T Y^{-1}(k)\\
&\times Z(k)E_iZ^T(k)Y^{-1}(k)F_i + 2\bar{M}(k)\bar{C}(k)P(k)\bar{C}^T(k)\bar{M}^T(k)\\
&\times Y^{-1}(k)Z(k)E_iZ^T(k)Y^{-1}(k)\Big\}.
\end{aligned}$$

Using the well-known matrix identity $\mathrm{tr}\{\Gamma\Delta\} = \mathrm{tr}\{\Delta\Gamma\}$, where Γ and Δ are appropriately dimensioned, we have

$$\begin{aligned}
\frac{\partial \mathrm{tr}\{P(k+1)\}}{\partial m(k)} &= \sum_{i=1}^n \frac{2}{m(k)} \mathrm{tr}\Big\{Y^{-1}(k)Z(k)E_iZ^T(k)Y^{-1}(k)\Big[F_iY(k)F_i^T\\
&- Y(k) - \bar{M}(k)\bar{C}(k)P(k)\bar{C}^T(k)\bar{M}^T(k)Y^{-1}(k)F_iY(k)F_i^T - F_i\\
&\times Y(k)F_i^T Y^{-1}(k)\bar{M}(k)\bar{C}(k)P(k)\bar{C}^T(k)\bar{M}^T(k) + F_i\bar{M}(k)\bar{C}(k)\\
&\times P(k)\bar{C}^T(k)\bar{M}^T(k)F_i^T + \bar{M}(k)\bar{C}(k)P(k)\bar{C}^T(k)\bar{M}^T(k)\Big]\Big\}\\
&= \sum_{i=1}^n \frac{2}{m(k)} \mathrm{tr}\Big\{Y^{-1}(k)Z(k)E_iZ^T(k)Y^{-1}(k)\Big\{-\Big[I - F_iY(k)\\
&\times F_i^T Y^{-1}(k)\Big]\Big[Y(k) - \bar{M}(k)\bar{C}(k)P(k)\bar{C}^T(k)\bar{M}^T(k)\Big]\Big[I - F_i\\
&\times Y(k)F_i^T Y^{-1}(k)\Big]^T - F_i\Big[Y(k) - \bar{M}(k)\bar{C}(k)P(k)\bar{C}^T(k)\\
&\times \bar{M}^T(k)\Big]F_i + F_iY(k)F_i^T Y^{-1}(k)[Y(k) - \bar{M}(k)\bar{C}(k)P(k)
\end{aligned}$$

$$\times \bar{C}^T(k)\bar{M}^T(k)\big]Y^{-1}(k)F_i^T Y(k)F_i\bigg\}\bigg\}. \tag{3.55}$$

Based on the fact that $\text{eig}(\Xi\Psi) = \text{eig}(\Psi\Xi)$, where Ξ and Ψ are square matrices with the same dimension and $\text{eig}(\cdot)$ denotes the eigenvalues of a matrix, we have that $\text{eig}(Y^{-1}(k)F_i Y(k)) = \text{eig}(F_i)$. Therefore, the eigenvalues of $Y^{-1}(k)F_i Y(k) - I$ are all either 0 or -1. Also notice that $Y(k) - \bar{M}(k)\bar{C}(k)P(k)\bar{C}^T(k)\bar{M}^T(k) = U(k) + V(k) > 0$. Then, it follows from (3.55) that,

$$\frac{\partial \text{tr}\{P(k+1)\}}{\partial m(k)} \leq \sum_{i=1}^{n}\frac{2}{m(k)}\text{tr}\bigg\{Y^{-1}(k)Z(k)E_i Z^T(k)Y^{-1}(k)\bigg\{-\big[I - F_i Y(k)$$
$$\times F_i^T Y^{-1}(k)\big]\big[Y(k) - \bar{M}(k)\bar{C}(k)P(k)\bar{C}^T(k)\bar{M}^T(k)\big]$$
$$\times \big[I - F_i Y(k)F_i^T Y^{-1}(k)\big]^T\bigg\}\bigg\}. \tag{3.56}$$

Denote

$$X_i(k) = -\big[I - F_i Y(k)F_i^T Y^{-1}(k)\big]\big[Y(k) - \bar{M}(k)\bar{C}(k)$$
$$\times P(k)\bar{C}^T(k)\bar{M}^T(k)\big]\big[I - F_i Y(k)F_i^T Y^{-1}(k)\big]^T.$$

It is obvious that $X_i(k) \leq 0$. Then, there always exists a matrix $\mathcal{X}_i(k)$ such that $X_i(k) = -\mathcal{X}_i(k)\mathcal{X}_i^T(k)$ based on the eigenvalue decomposition of $X_i(k)$, and (3.56) can be written as

$$\frac{\partial \text{tr}\{P(k+1)\}}{\partial m(k)} \leq -\sum_{i=1}^{n}\frac{2}{m(k)}\text{tr}\big\{Y^{-1}(k)Z(k)E_i Z^T(k)Y^{-1}(k)\mathcal{X}_i(k)\mathcal{X}_i^T(k)\big\}$$
$$= -\sum_{i=1}^{n}\frac{2}{m(k)}\text{tr}\big\{\mathcal{X}_i^T(k)Y^{-1}(k)Z(k)E_i Z^T(k)Y^{-1}(k)\mathcal{X}_i(k)\big\}. \tag{3.57}$$

Noticing that the trace of a symmetric positive semidefinite matrix is always nonnegative, for all $m(k) \in (0,1)$, we have

$$\frac{\partial \text{tr}\{P(k+1)\}}{\partial m(k)} \leq 0, \tag{3.58}$$

which means that $\text{tr}\{P(k+1)\}$ is nonincreasing as $m(k)$ increases. The proof of this theorem is now complete.

Main Results

Remark 3.6 *The finding in Theorem 3.3 is in accordance with the intuition. Actually, the increase of $m(k)$ can be interpreted that the sensor or the communication channel is under a better working condition. Theorem 3.3 shows that the filtering performance gets improved when more information about the measurements is received at each sensor node.*

To further illustrate the engineering significance of Theorem 3.3, let us now look at the two interesting extreme cases of $F_i = I$ and $F_i = 0$.

In case of $F_i = I$, each node just estimates the dynamics based on $\hat{x}_i(k+1) = A(k)\hat{x}_i(k)$ (without any measurement signals). In such a situation, the measurements $y_i(k)$ have no effect on the filtering performance, and thus $\partial P(k+1)/\partial m_i(k) = 0$ for all $1 \leq i \leq n$.

In case of $F_i = 0$, we consider each sensor to be subject to individual sensor gain degradation, that is, $\mathbb{E}\{\delta_i(k)\} = m_i(k)$. The following result is easily accessible from Theorem 3.3, and therefore only a sketch of the proof is provided.

Corollary 3.2 *If the sensor topology is completely connected, then $\mathrm{tr}\{P(k+1)\}$ is nonincreasing as $m_i(k)$ increases for all $1 \leq i \leq n$.*

Proof *When the directed graph is completely connected, $F_i = 0$ for all $1 \leq i \leq n$ according to its definition. From (3.53) we have:*

$$\frac{\partial \mathrm{tr}\{P(k+1)\}}{\partial m_i(k)} = -\frac{\partial}{\partial m_i(k)}\mathrm{tr}\{Z^T(k)Y^{-1}(k)Z(k)\}$$

$$= \mathrm{tr}\bigg\{2E_i\bar{C}(k)P(k)\bar{C}^T(k)\bar{M}^T(k)Y^{-1}(k)Z(k)Z^T(k)Y^{-1}(k)$$

$$- 2E_i\bar{C}(k)P(k)\bar{A}^T(k)Z^T(k)Y^{-1}(k)\bigg\}$$

$$= \mathrm{tr}\bigg\{2E_i\bar{C}(k)P(k)\big[\bar{C}^T(k)\bar{M}^T(k)Y^{-1}(k)\bar{M}(k)\bar{C}(k)$$

$$- P^{-1}(k)\big]P(k)\bar{A}^T(k)\bar{A}(k)P(k)\bar{C}^T(k)\bar{M}^T(k)Y^{-1}(k)\bigg\}.$$

(3.59)

It follows from (3.13) that $\bar{C}^T(k)\bar{M}^T(k)Y^{-1}(k)\bar{M}(k)\bar{C}(k) < P^{-1}(k)$ and, subsequently,

$$\frac{\partial \mathrm{tr}\{P(k+1)\}}{\partial m_i(k)} \leq 0, \qquad (3.60)$$

which concludes the proof.

Remark 3.7 *In the main results of this chapter, a minimum-variance filter is designed at each time step even if the sensor network is not completely* connected, and the filter is characterized by resorting to the statistical information on the sensor gain degradation. Then, sufficient conditions are established under which the estimation error is exponentially bounded in mean square. Furthermore, the relationship between the filter performance (in terms of the estimation variance) and the statistical law of the sensor gain degradation (in terms of the degradation probability) is revealed and analyzed in a mathematically rigorous way. The matrices T_i and F_i reflect the effects of the possibly sparse topology in the filter design and performance analysis, which means that the topology is taken into account and the proposed approach is applicable in distributed settings. The matrices T_i and F_i quantify the effects of the sparse topology of the sensor network in the filter design and performance analysis, which means that the topology is taken into account and the proposed approach is applicable in distributed settings. The main differences between our work and the decentralized multisensor Kalman filter in [135] are threefold: 1) the consideration of the sparse sensor network; 2) the investigation on stochastic sensor gain degradation phenomenon, and 3) the estimation performance analysis with respect to the boundedness and the monotonicity.

In next section, a numerical example is provided to illustrate our proposed filter design scheme.

3.3 Numerical Example

The time-varying target plant is modeled by (3.1) with the following parameters:

$$A(k) = \begin{bmatrix} 0.1315 + 0.0054\sin(k) & 0.0537 \\ 0.0201 & -0.1007 \end{bmatrix}, \quad \tilde{A}(k) = I.$$

Suppose that $w(k)$ is a zero-mean Gaussian white noise with covariance $2.5 \times 10^{-5}I$, and $\rho(k)$ uniformly distributes over $[-0.001, 0.001]$. The initial value of the state $x(0)$ is uniformly distributed over $[-0.1, 0]$, and therefore $\mathbb{E}\{x(0)\} = [-0.05, -0.05]^T$.

The sensor network is represented by a directed graph $\mathcal{G} = (\mathcal{V}, \mathcal{E}, \mathcal{A})$ with the set of nodes $\mathcal{V} = \{1, 2, 3, 4\}$, the set of edges $\mathcal{E} = \{(1,1), (1,2), (2,1), (2,2), (2,3), (3,1), (3,3), (4,1), (4,4)\}$, and the adjacency elements associated with the edges of the graph are $a_{ij} = 1$.

Numerical Example

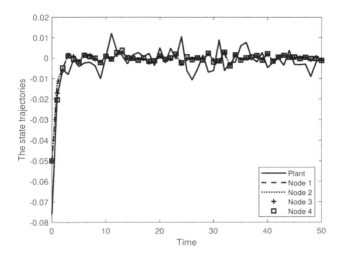

FIGURE 3.1: The state x_1 and its estimate

The dynamics of the sensor nodes are described by (3.2) with parameters as follows:

$$C_1(k) = [0.82, \ 0.62], \quad C_2(k) = [0.75, \ 0.80],$$
$$C_3(k) = [0.74, \ 0.75], \quad C_4(k) = [0.75, \ 0.70].$$

The additive noises $v_i(k)$ are uncorrelated Gaussian sequences whose covariances are $2.5 \times 10^{-5} I$. $\delta_i(k)$ is uniformly distributed, respectively, over $[0.45 + 0.1i, \ 0.95 + 0.1i]$ for $i = 1, \ldots, 4$.

Since the sensor network topology is not completely connected, the results of Theorem 1, Part b) are adopted. At each node, $P(k)$ is updated with (3.19) and $H_{ij}(k)$ is determined with (3.18). Fig. 3.1 and Fig. 3.2 depict the trajectories for the states and estimates, respectively. Fig. 3.3 plots the real estimation errors and the bound calculated from Theorem 3.2. In Fig. 3.4, we assume that all the sensor gain degradations have the same statistics, i.e., variables $\delta_i(k)$ ($i = 1, \ldots, 4$) are uniformly distributed over the same interval with the length of 0.5. The relationship between the accumulative estimation error variance in 50 time steps and the value of $m(k)$ is shown in Fig. 3.4. As pointed out in Theorem 3.3, the filtering error variance decreases when $m(k)$ increases.

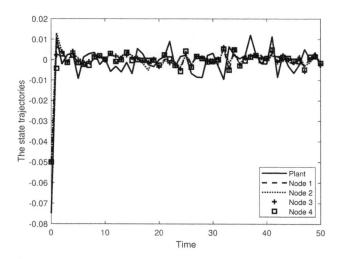

FIGURE 3.2: The state x_2 and its estimate

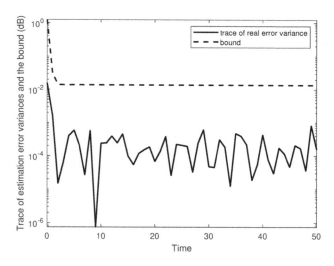

FIGURE 3.3: The estimation error and bound

Conclusions

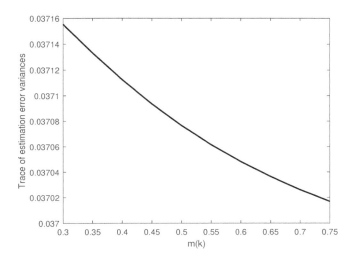

FIGURE 3.4: The estimation error and sensor gain degradation

3.4 Conclusions

In this chapter, the minimum variance filtering problem has been investigated for a class of time-varying systems through sensor networks subject to both additive/multiplicative noises and stochastic sensor gain degradation. The gain degradation phenomenon for each individual sensor occurs in a random way governed by a random variable distributed over the interval $[0, 1]$. The distributed filter has been designed recursively by solving a set of Riccati-like matrix equations, such that the overall estimation error variance is minimized at each time step. The performance of the designed filters has been analyzed in terms of both the boundedness and monotonicity. Specifically, sufficient conditions have been obtained under which the estimation error is exponentially bounded in mean square. Moreover, the monotonicity property for the error variance with respect to the sensor gain degradation has been thoroughly discussed in the case where all the sensors are subject to gain degradation with the same probability. Numerical simulations have been exploited to show the effectiveness of the proposed filtering algorithm.

4

H_∞ Filtering for Nonlinear Systems with Stochastic Sensor Saturations and Markov Time Delays

In the past decades, research on the filtering problem has been the mainstream of control discipline due to its significance in signal processing, communication, navigation and tracking, finance, etc. Fruitful results have been reported in this area and many effective methodologies have been proposed. Traditional Kalman filter and extended Kalman filter can solve the estimation problem, respectively, for linear and nonlinear systems in the least mean square sense [16]. The H_∞ filtering has been extensively studied to guarantee that the L_2 gain from the disturbance to the estimation error is less than a predefined positive level [60]. For stochastic systems, the concept of stability in probability is important because it can describe the system dynamics in a probabilistic way. So far, the filtering problem in the sense of stability in probability has stirred some initial research interests.

In practical engineering, sensors cannot generate measured outputs with unbounded amplitudes due to physical and technological limitations such as nonlinear transition shift sensors, position sensors, displacement sensors, pressure sensors, and so on. This phenomenon, often referred to as sensor saturation, would bring in extra challenges to the filtering problem. So far, the filtering problem with sensor saturations has drawn much research attention [35]. In most of the reported literature, the sensor saturations have been assumed to occur definitely and the sector-bounded conditions have been used to formulate the saturation-induced errors. Nevertheless, sensors in practical systems might be subject to some transient phenomena especially in changeable environments such as power grids [57, 82]. In these cases, the saturations may occur in a random way owing to various reasons such as random sensor failures and abrupt environmental changes [164]. The filtering problem with stochastic sensor saturations has not received adequate research attention yet, not to mention the case where the stability in probability is also taken into account to quantify the performance. As such, it would be interesting to examine how the saturation levels and the statistical characteristics of the sensor saturations would influence the stability in probability for the filter design problem.

It is well known that time delays widely exist in many practical systems and may deteriorate the system performances if not appropriately coped with. The filtering problem with time delays has been investigated for a variety of

DOI: 10.1201/9781003309482-4

systems such as two-dimensional systems and neural networks. A widely adopted way to formulate the stochastic time delays is the finite state Markov chain method, which can characterize the relationship between the delays at different time steps. An H_∞ filter has been proposed in [94] for nonlinear systems with model uncertainties and Markov delays. Also, least-squares estimators for systems with Markov delays have been designed recursively. Nevertheless, when the Markov time delay issue is coupled with stochastic sensor saturations, the filtering problem for discrete nonlinear systems with guaranteed stability in probability still remains as an ongoing research issue. In fact, it is non-trivial to establish a unified framework to accommodate nonlinearities, stochastic sensor saturations as well as Markov time delays simultaneously. The main purpose of this chapter is to shorten such a gap.

In this chapter, we aim to address filtering problem, in the sense of asymptotic stability in probability, for a class of nonlinear systems with stochastic sensor saturations and Markov time delays. A Bernoulli-distributed sequence is employed to regulate the stochastic sensor saturations. Time delays in the measurement transmissions are governed by a discrete Markov chain with finite states. Sufficient conditions are established to guarantee the desired stability in probability and determine the filter parameters. The linear filter is further investigated as a special case. Two simulation examples are presented to show the effectiveness of the proposed method. The main novelty of the chapter lies in the following aspects: 1) a comprehensive model is established which covers nonlinearities, stochastic sensor saturations, and Markov time delays; 2) sufficient conditions are achieved under which the designed filter is asymptotically stable in probability in the disturbance-free case and also robust to exogenous disturbances under the zero-initial condition; and 3) the conditions are specialized to some linear filters such that the simplified results are more applicable in practice.

4.1 Problem Formulation

Consider the following class of stochastic discrete-time nonlinear systems:

$$\begin{cases} x_{k+1} = f(x_k) + g(x_k)v_k + (h(x_k) + s(x_k)v_k)w_k, \\ z_k = m(x_k), \\ \tilde{y}_k = v_k \sigma(l(x_k)) + (1 - v_k)l(x_k) + k(x_k)v_k, \\ y_k = \tilde{y}_{k-d_k}, \end{cases} \quad (4.1)$$

where $x_k \in \mathbb{R}^{n_x}$ is the state; $z_k \in \mathbb{R}^{n_z}$ is the signal to be estimated; w_k is a one-dimensional and zero-mean Gaussian white noise sequence with $\mathbb{E}\{w_k^2\} = r^2$ (r is a known positive scalar); $v_k \in \mathbb{R}^{n_v}$ is the exogenous disturbance satisfying $\{v_k\}_{k\in\mathbb{N}} \in l_2([0, +\infty), \mathbb{R}^{n_v})$; $\tilde{y}_k \in \mathbb{R}^{n_y}$ is the measurement before transmission; $y_k \in \mathbb{R}^{n_y}$ is the received signal impaired by communication delays; d_k is the

Problem Formulation

homogeneous discrete-time Markov chain defined on $N \triangleq \{0, 1, \cdots, d-1\}$ with the one-step transition matrix $(\pi_{ij})_{d \times d}$ and the initial distribution π_0, where $d > 0$ is a fixed integer.

The nonlinear functions $f : \mathbb{R}^{n_x} \to \mathbb{R}^{n_x}$, $g : \mathbb{R}^{n_x} \to \mathbb{R}^{n_x \times n_v}$, $h : \mathbb{R}^{n_x} \to \mathbb{R}^{n_x}$, $s : \mathbb{R}^{n_x} \to \mathbb{R}^{n_x \times n_v}$, $m : \mathbb{R}^{n_x} \to \mathbb{R}^{n_z}$, $l : \mathbb{R}^{n_x} \to \mathbb{R}^{n_y}$, and $k : \mathbb{R}^{n_x} \to \mathbb{R}^{n_y \times n_v}$ are assumed to be smooth, matrix-valued functions with $f(0) = 0$, $h(0) = 0$, $m(0) = 0$, and $l(0) = 0$. $v_k \in \mathbb{R}$ is a Bernoulli distributed white sequence taking values on 0 or 1 with

$$\begin{cases} \text{Prob}\{v_k = 1\} = \bar{v}, \\ \text{Prob}\{v_k = 0\} = 1 - \bar{v}, \end{cases} \quad (4.2)$$

where $\bar{v} \in [0, 1]$ is a known scalar. For a vector $q = \begin{bmatrix} q_1, \ldots, q_{n_y} \end{bmatrix}^T$, the saturation function $\sigma : \mathbb{R}^{n_y} \to \mathbb{R}^{n_y}$ is defined as:

$$\sigma(q) = \begin{bmatrix} \sigma_1(q_1), \cdots, \sigma_{n_y}(q_{n_y}) \end{bmatrix}^T, \quad (4.3)$$

with $\sigma_i(q_i) = \text{sign}(q_i)\min(q_{i,\max}, |q_i|)$, where the notation of "sign" denotes the signum function and $q_{i,\max} > 0$ denotes the ith saturation level.

For notational brevity, we set

$$\bar{x}_k = \begin{bmatrix} x_k^T, \cdots, x_{k-(d-1)}^T \end{bmatrix}^T, \bar{v}_k = \begin{bmatrix} v_k^T, \cdots, v_{k-(d-1)}^T \end{bmatrix}^T,$$

$$\bar{f}(\bar{x}_k) = \begin{bmatrix} f^T(x_k), x_k^T, \cdots, x_{k-(d-2)}^T \end{bmatrix}^T, \bar{g}(\bar{x}_k) = \text{diag}\{g(x_k), 0, \cdots, 0\},$$

$$\bar{h}(\bar{x}_k) = \begin{bmatrix} h^T(x_k), 0, \cdots, 0 \end{bmatrix}^T, \bar{s}(\bar{x}_k) = \text{diag}\{s(x_k), 0, \cdots, 0\},$$

$$\bar{y}_k = \begin{bmatrix} \tilde{y}_k^T, \cdots, \tilde{y}_{k-(d-1)}^T \end{bmatrix}^T, \Upsilon_k = \text{diag}\{v_k, \cdots, v_{k-(d-1)}\} \otimes I_{n_y},$$

$$\bar{m}(\bar{x}_k) = m(x_k), \bar{l}(\bar{x}_k) = \begin{bmatrix} l^T(x_k), \cdots, l^T(x_{k-(d-1)}) \end{bmatrix}^T,$$

$$\bar{\sigma}(\bar{l}(\bar{x}_k)) = \begin{bmatrix} \sigma^T(l(x_k)), \cdots, \sigma^T(l(x_{k-(d-1)})) \end{bmatrix}^T,$$

$$\bar{k}(\bar{x}_k) = \text{diag}\{k(x_k), \cdots, k(x_{k-(d-1)})\},$$

and $C_{d_k} = [0, \cdots, 0, I, 0, \cdots, 0]$ with the $(d_k + 1)$th block being an identity.

With the defined notations, the original model (4.1) can be written in the following form:

$$\begin{cases} \bar{x}_{k+1} = \bar{f}(\bar{x}_k) + \bar{g}(\bar{x}_k)\bar{v}_k + (\bar{h}(\bar{x}_k) + \bar{s}(\bar{x}_k)\bar{v}_k)w_k, \\ z_k = \bar{m}(\bar{x}_k), \\ \bar{y}_k = \Upsilon_k \bar{\sigma}(\bar{l}(\bar{x}_k)) + (I - \Upsilon_k)\bar{l}(\bar{x}_k) + \bar{k}(\bar{x}_k)\bar{v}_k, \\ y_k = C_{d_k}\bar{y}_k. \end{cases} \quad (4.4)$$

Remark 4.1 *The measurement equations in (4.1) can effectively the Markov transmission delays and the stochastic sensor saturations. The random variables Υ_k and d_k account for the sensor saturations and time delays, respectively. The statistics of Υ_k and d_k would be utilized in the filter design procedure.*

In this chapter, a full-order filter of the following structure is adopted:

$$\begin{cases} \hat{x}_{k+1} = \hat{f}(\hat{x}_k) + \hat{g}(\hat{x}_k, d_k)y_k, \\ \hat{z}_k = \hat{m}(\hat{x}_k), \end{cases} \quad (4.5)$$

where $\hat{x}_k \in \mathbb{R}^{d \times n_x}$ is the state estimate; $\hat{z}_k \in \mathbb{R}^{n_z}$ is the estimate of z_k, and $\hat{f}$, $\hat{g}$, and $\hat{m}$ are filter parameters of appropriate dimensions that are to be determined with $f(0) = 0$, $m(0) = 0$, and $\hat{x}_0 = 0$.

By introducing a new vector $\eta_k = \left[\bar{x}_k^T, \hat{x}_k^T\right]^T$ and letting the filtering error be $\tilde{z}_k = z_k - \hat{z}_k$, an augmented system is obtained as follows:

$$\begin{cases} \eta_{k+1} = \tilde{f}(\eta_k, d_k) + \tilde{g}(\eta_k, d_k)v_k + (\tilde{h}(\eta_k) + \tilde{s}(\eta_k)\bar{v}_k)w_k, \\ \tilde{z}_k = \bar{m}(\bar{x}_k) - \hat{m}(\hat{x}_k), \end{cases} \quad (4.6)$$

where

$$\tilde{f}(\eta_k, d_k) = \begin{bmatrix} \bar{f}(\bar{x}_k) \\ \hat{f}(\hat{x}_k) + \hat{g}(\hat{x}_k, d_k)C_{d_k}\Upsilon_k\bar{\sigma}(\bar{l}(\bar{x}_k)) + \hat{g}(\hat{x}_k, d_k)C_{d_k}(I - \Upsilon_k)\bar{l}(\bar{x}_k) \end{bmatrix},$$

$$\tilde{g}(\eta_k, d_k) = \begin{bmatrix} \bar{g}(\bar{x}_k) \\ \hat{g}(\hat{x}_k, d_k)C_{d_k}\bar{k}(\bar{x}_k) \end{bmatrix}, \tilde{h}(\eta_k) = \begin{bmatrix} \bar{h}(\bar{x}_k) \\ 0 \end{bmatrix}, \tilde{s}(\eta_k) = \begin{bmatrix} \bar{s}(\bar{x}_k) \\ 0 \end{bmatrix}.$$

Before proceeding, let us first introduce the following definition, which is a discrete version of that in [81].

Definition 4.1 *The solution $\eta_k = 0$ of (4.6) is said to be:*

1. *Stable in probability if for every pair of $\epsilon \in (0,1)$ and $\alpha > 0$, there exists a $\delta = \delta(\epsilon, \alpha) > 0$ such that*

$$\text{Prob}\{\|\eta_k\| < \alpha, \forall k \in \mathbb{N}\} \geq 1 - \epsilon, \quad \text{whenever } \|\eta_0\| < \delta.$$

2. *Asymptotically stable in probability if the origin of (4.1) is stable in probability, and*

$$\lim_{k \to \infty} \text{Prob}\{\|\eta_k\| = 0\} = 1, \quad \forall \eta_0 \in \mathbb{R}^{2n_x}.$$

In this chapter, we aim to design the filter parameters $\hat{f}(\hat{x}_k)$, $\hat{g}(\hat{x}_k, d_k)$ and $\hat{m}(\hat{x}_k)$ in (4.5) such that the following requirements are simultaneously satisfied:

1. The zero-solution of the augmented system (4.6) with $\bar{v}_k = 0$ is asymptotically stable in probability.

2. Under the zero-initial condition, the filtering error $\tilde{z}_k$ satisfies

$$\sum_{k=0}^{\infty} \mathbb{E}\left\{\|\tilde{z}_k\|^2\right\} < \gamma^2 \sum_{k=0}^{\infty} \mathbb{E}\left\{\|\bar{v}_k\|^2\right\} \quad (4.7)$$

for all nonzero $\bar{v}_k$, where $\gamma > 0$ is a given disturbance attenuation level.

Main Results 65

The filtering problem for the addressed nonlinear system will be solved in the next section, and the results will be applied to some special cases for practical convenience.

4.2 Main Results

We start with the following definitions and lemma which will be used in the development of the main results.

Definition 4.2 *[56] A function $V : \mathbb{R}^n \to \mathbb{R}_+$ is said to be positive definite if $V(0) = 0$ and $V(x) > 0$ for all $x \in \mathbb{R}^n \backslash \{0\}$.*

Definition 4.3 *[79] A function $\kappa : \mathbb{R}_+ \to \mathbb{R}_+$ is said to be a $\mathcal{K}$ class function if it is continuous, strictly increasing and $\kappa(0) = 0$. A $\mathcal{K}$ class function $\kappa(\cdot)$ is said to be a $\mathcal{K}_\infty$ class function if $\kappa(r) \to +\infty$ as $r \to +\infty$.*

Lemma 4.1 *If there exists a positive definite function $V : \mathbb{R}^{2d \times n_x} \to \mathbb{R}_+$, two $\mathcal{K}_\infty$ class functions κ_1, κ_2, and a $\mathcal{K}$ class function κ_3 such that for all $k \in \mathbb{N}$,*

$$\kappa_1(\|\eta\|) \leq V(\eta) \leq \kappa_2(\|\eta\|), \forall \eta \in \mathbb{R}^{2d \times n_x}, \quad (4.8)$$

$$\mathbb{E}\{V(\eta_{k+1})|\eta_k\} - V(\eta_k) \leq -\kappa_3(\|\eta_k\|), \quad (4.9)$$

then the origin of (4.6) with $\bar{v}_k = 0$ is asymptotically stable in probability.

Proof Based on (4.9) and the definitions of $\mathcal{K}$ class functions and positive definite functions, we have

$$0 \leq \mathbb{E}\{V(\eta_{k+1})\} \leq \mathbb{E}\{V(\eta_k)\} \leq \cdots V(\eta_0). \quad (4.10)$$

Then, for any $\alpha > 0$ and $k \in \mathbb{N}$, we have

$$\text{Prob}\{\|\eta_k\| \geq \alpha\} = \text{Prob}\{\kappa_1(\|\eta_k\|) \geq \kappa_1(\alpha)\} \leq \text{Prob}\{V(\eta_k) \geq \kappa_1(\alpha)\}. \quad (4.11)$$

Since $V(\eta_k)$ is nonnegative, we have

$$\mathbb{E}\{V(\eta_k)\} \geq \kappa_1(\alpha)\text{Prob}\{V(\eta_k) \geq \kappa_1(\alpha)\}. \quad (4.12)$$

With (4.11) and (4.12), it can be obtained that

$$\text{Prob}\{V(\eta_k) \geq \kappa_1(\alpha))\} \leq \frac{\mathbb{E}\{V(\eta_k)\}}{\kappa_1(\alpha))} \leq \frac{V(\eta_0)}{\kappa_1(\alpha))} \leq \frac{\kappa_2(\|\eta_0\|)}{\kappa_1(\alpha)}. \quad (4.13)$$

Therefore, for any $\epsilon > 0$, we can choose $\delta = \kappa_2^{-1}(\epsilon \kappa_1(\alpha)) > 0$ such that

$$\text{Prob}\{\|\eta_k\| < \alpha, \forall k \in \mathbb{N}\} \geq 1 - \epsilon, \text{ whenever } \|\eta_0\| < \delta.$$

With (4.10), we can also find a $V_\infty \geq 0$ such that
$$\lim_{k\to\infty} \mathbb{E}\{V(\eta_k)\} = V_\infty. \tag{4.14}$$

Substituting (4.14) into (4.9) yields
$$\lim_{k\to\infty} \mathbb{E}\{\kappa_3(\|\eta_k\|)\} = 0, \tag{4.15}$$

which implies that
$$\lim_{k\to\infty} \text{Prob}\{\|\eta_k\| = 0\} = 1.$$

The proof is now complete.

With Lemma 4.1, sufficient conditions are going to be established in the following theorem to facilitate the filter design.

Theorem 4.1 *Given a disturbance attenuation level $\gamma > 0$. If there exist two positive definite matrices $P = P^T > 0$ and $Q = Q^T > 0$, and a $\mathcal{K}$ class function κ_3 satisfying the inequalities*

$$\begin{cases} \mathbb{H}(\bar{x},\hat{x},\tau) = B(\bar{x},\hat{x},\tau)A^{-1}(\bar{x},\hat{x},\tau)B^T(\bar{x},\hat{x},\tau) + r^2\bar{h}^T(\bar{x})P\bar{h}(\bar{x}) + D(\bar{x},\hat{x},\tau) \\ \qquad + \|\tilde{z}\|^2 + \kappa_3(\|\eta\|) < 0, \text{ for any } \bar{x}, \hat{x} \in \mathbb{R}^{d \times n_x}, \tau \in N, \\ A(\bar{x},\hat{x},\tau) > 0, \text{ for any } \bar{x}, \hat{x} \in \mathbb{R}^{d \times n_x}, \tau \in N, \end{cases} \tag{4.16}$$

where

$$\bar{\Upsilon} = \bar{v}I_{d\times n_y}, \quad \bar{\Upsilon}_2 = \begin{bmatrix} 1-\bar{v} & (1-\bar{v})^2 & \cdots & (1-\bar{v})^2 \\ * & 1-\bar{v} & \cdots & \vdots \\ \vdots & \vdots & \ddots & \vdots \\ * & \cdots & \cdots & 1-\bar{v} \end{bmatrix} \otimes \mathbf{1}_{n_y \times n_y},$$

$$\bar{\Upsilon}_1 = \begin{bmatrix} \bar{v} & \bar{v}^2 & \cdots & \bar{v}^2 \\ * & \bar{v} & \cdots & \vdots \\ \vdots & \vdots & \ddots & \vdots \\ * & \cdots & \cdots & \bar{v} \end{bmatrix} \otimes \mathbf{1}_{n_y \times n_y}, \quad \bar{\Upsilon}_3 = \begin{bmatrix} 0 & \bar{v}(1-\bar{v}) & \cdots & \bar{v}(1-\bar{v}) \\ * & 0 & \cdots & \vdots \\ \vdots & \vdots & \ddots & \vdots \\ * & \cdots & \cdots & 0 \end{bmatrix} \otimes \mathbf{1}_{n_y \times n_y},$$

$$A(\bar{x},\hat{x},\tau) = \gamma^2 I - r^2 \bar{s}^T(\bar{x})P\bar{s}(\bar{x}) - \bar{g}^T(\bar{x})P\bar{g}(\bar{x}) - \bar{k}^T(\bar{x})S_1(\bar{x},\hat{x},\tau)\bar{k}(\bar{x}), \tag{4.17}$$

$$B(\bar{x},\hat{x},\tau) = r^2 \bar{h}^T(\bar{x})P\bar{s}(\bar{x}) + \bar{f}^T(\bar{x})P\bar{g}(\bar{x}) + \hat{f}^T(\hat{x})Q\hat{g}(\hat{x},\tau)C_\tau \bar{k}(\bar{x}) \\ + S_2^T(\bar{x},\hat{x},\tau)Q\hat{g}(\hat{x},\tau)C_\tau \bar{k}(\bar{x}), \tag{4.18}$$

$$D(\bar{x},\hat{x},\tau) = (\bar{f}(\bar{x})+\bar{x})^T P(\bar{f}(\bar{x})-\bar{x}) + (\hat{f}(\hat{x})+\hat{x})^T Q(\hat{f}(\hat{x})-\hat{x}) \\ + 2\hat{f}^T(\hat{x})QS_2(\bar{x},\hat{x},\tau) + \bar{\sigma}^T(\bar{l}(\bar{x}))(\bar{\Upsilon}_1 \circ S_1(\hat{x},\tau))\bar{\sigma}(\bar{l}(\bar{x})) \\ + 2\bar{\sigma}^T(\bar{l}(\bar{x}))(\bar{\Upsilon}_3 \circ S_1(\hat{x},\tau))\bar{l}(\bar{x}) + \bar{l}^T(\bar{x})(\bar{\Upsilon}_2 \circ S_1(\hat{x},\tau))\bar{l}(\bar{x}), \tag{4.19}$$

$$S_1(\hat{x},\tau) = C_\tau^T \hat{g}^T(\hat{x},\tau)Q\hat{g}(\hat{x},\tau)C_\tau, \tag{4.20}$$

$$S_2(\bar{x},\hat{x},\tau) = \hat{g}(\hat{x},\tau)C_\tau \bar{\Upsilon}\bar{\sigma}(\bar{l}(\bar{x})) + \hat{g}(\hat{x},\tau)C_\tau (I-\bar{\Upsilon})\bar{l}(\bar{x}), \tag{4.21}$$

Main Results

then the filtering problem for system (4.1) is solved by (4.5) in the sense of asymptotic stability in probability.

Proof Set $V^{(1)}(\bar{x}) = \bar{x}^T P \bar{x}$ and $V^{(2)}(\hat{x}) = \hat{x}^T Q \hat{x}$. Define

$$\bar{\rho} = \max\{\upsilon_{\max}(P), \upsilon_{\max}(Q)\}, \quad \underline{\rho} = \min\{\upsilon_{\min}(P), \upsilon_{\min}(Q)\},$$

where $\upsilon_{\max}(\cdot)$ and $\upsilon_{\max}(\cdot)$ denote the maximum and the minimum eigenvalue of the square matrix, respectively. Then, two $\mathcal{K}_\infty$ functions can be defined as $\kappa_1(\|\eta\|) = \underline{\rho}\|\eta\|^2$ and $\kappa_2(\|\eta\|) = \bar{\rho}\|\eta\|^2$, and it follows that $\kappa_1(\|\eta\|) \leq V(\eta) \leq \kappa_2(\|\eta\|)$.

With $\mathbb{E}\{w_k\} = 0$, $\mathbb{E}\{w_k^2\} = r^2$ and (4.6), it can be obtained that

$$\mathbb{E}\{V(\eta_{k+1})|\eta_k\} - V(\eta_k) + \mathbb{E}\{\|\tilde{z}_k\|^2\} - \gamma^2 \mathbb{E}\{\|\bar{v}_k\|^2\} + \kappa_3(\|\eta_k\|)$$
$$= \mathbb{E}\Big\{\bar{f}^T(\bar{x}_k)P\bar{f}(\bar{x}_k) + \big(\hat{f}(\hat{x}_k) + \hat{g}(\hat{x}_k, d_k)C_{d_k}\bar{\Upsilon}\bar{\sigma}(\bar{l}(\bar{x}_k)) + \hat{g}(\hat{x}_k, d_k)C_{d_k}(I$$
$$- \bar{\Upsilon})\bar{l}(\bar{x}_k)\big)^T Q\big(\hat{f}(\hat{x}_k) + \hat{g}(\hat{x}_k, d_k)C_{d_k}\bar{\Upsilon}\bar{\sigma}(\bar{l}(\bar{x}_k)) + \hat{g}(\hat{x}_k, d_k)C_{d_k}(I - \bar{\Upsilon})\bar{l}(\bar{x}_k)\big)$$
$$+ r^2\Big(\bar{v}_k^T s^T(\bar{x}_k)Ps(\bar{x}_k)\bar{v}_k + \bar{h}^T(\bar{x}_k) \times P\bar{h}^T(\bar{x}_k) + 2\bar{h}^T(\bar{x}_k)Ps(\bar{x}_k)\bar{v}_k\Big)$$
$$+ \bar{v}_k^T\big(\bar{g}^T(\bar{x})P\bar{g}(\bar{x}) + \bar{k}^T(\bar{x})C_\tau^T \hat{g}^T(\hat{x},\tau)Q\hat{g}(\hat{x},\tau)C_\tau \bar{k}(\bar{x})\big)\bar{v}_k$$
$$+ 2\bar{f}^T(\bar{x}_k)P\bar{g}(\bar{x}_k)\bar{v}_k + 2\big(\hat{f}(\hat{x}_k) + \hat{g}(\hat{x}_k, d_k)C_{d_k}\bar{\Upsilon}\bar{\sigma}(\bar{l}(\bar{x}_k))$$
$$+ \hat{g}(\hat{x}_k, d_k)C_{d_k}(I - \bar{\Upsilon})\bar{l}(\bar{x}_k)\big)^T Q \times \hat{g}(\hat{x},\tau)C_\tau \bar{k}(\bar{x})\bar{v}_k\Big\}$$
$$- \bar{x}_k^T P\bar{x}_k - \hat{x}_k^T Q\hat{x}_k + \mathbb{E}\{\|\tilde{z}_k\|^2\} - \gamma^2 \mathbb{E}\{\|\bar{v}_k\|^2\} + \kappa_3(\|\eta_k\|).$$

Completing the squares with respect to $\bar{v}_k$ yields that

$$\mathbb{E}\{V(\eta_{k+1})|\eta_k\} - V(\eta_k) + \mathbb{E}\{\|\tilde{z}_k\|^2\} - \gamma^2 \mathbb{E}\{\|\bar{v}_k\|^2\} + \kappa_3(\|\eta_k\|)$$
$$= \mathbb{E}\Big\{-(\bar{v}_k - \bar{v}_k^*)^T A(\bar{x}_k, \hat{x}_k, d_k)(\bar{v}_k - \bar{v}_k^*) + B(\bar{x}_k, \hat{x}_k, d_k)A^{-1}(\bar{x}_k, \hat{x}_k, d_k)$$
$$\times B^T(\bar{x}_k, \hat{x}_k, d_k) + D(\bar{x}_k, \hat{x}_k, d_k) + r^2 \bar{h}^T(\bar{x}_k)P\bar{h}(\bar{x}_k) + \|\tilde{z}_k\|^2 + \kappa_3(\|\eta_k\|)\Big\},$$

where $\bar{v}_k^* = A^{-1}(\eta_k, \eta_{\alpha_k}, d_k)B^T(\eta_k, \eta_{\alpha_k}, d_k)$. Based on (4.16), we have

$$\mathbb{E}\{V(\eta_{k+1})|\eta_k\} - V(\eta_k) + \mathbb{E}\{\|\tilde{z}_k\|^2\} - \gamma^2 \mathbb{E}\{\|\bar{v}_k\|^2\} + \kappa_3(\|\eta_k\|)$$
$$\leq B(\bar{x}_k, \hat{x}_k, d_k)A^{-1}(\bar{x}_k, \hat{x}_k, d_k)B^T(\bar{x}_k, \hat{x}_k, d_k) + D(\bar{x}_k, \hat{x}_k, d_k)$$
$$+ r^2 \bar{h}^T(\bar{x}_k)P\bar{h}(\bar{x}_k) + \|\tilde{z}_k\|^2 + \kappa_3(\|\eta_k\|)$$
$$= \mathbb{H}(\bar{x}_k, \hat{x}_k, d_k) < 0.$$

Now we have proved that

$$\mathbb{E}\{V(\eta_{k+1})|\eta_k\} - V(\eta_k) + \mathbb{E}\{\|\tilde{z}_k\|^2\} - \gamma^2 \mathbb{E}\{\|\bar{v}_k\|^2\} + \kappa_3(\|\eta_k\|) < 0. \quad (4.22)$$

Noticing that $\kappa_3(\|\eta_k\|) > 0$ and summing up (4.22) from 0 to positive integer N with respect to k, we have

$$\sum_{k=0}^{N} \mathbb{E}\{\|\tilde{z}_k\|^2\} < \gamma^2 \sum_{k=0}^{N} \mathbb{E}\{\|\bar{v}_k\|^2\} + \mathbb{E}\{V(0)\} - \mathbb{E}\{V(\eta_{N+1})\}. \quad (4.23)$$

Considering $V(\eta_{N+1}) > 0$, $V(0) = 0$ under the zero-initial condition and letting $N \to \infty$, we can get

$$\sum_{k=0}^{\infty} \mathbb{E}\{\|\tilde{z}_k\|^2\} < \gamma^2 \sum_{k=0}^{\infty} \mathbb{E}\{\|\bar{v}_k\|^2\}, \quad (4.24)$$

which means that the desired H_∞ performance requirement is met.

When $\bar{v}_k = 0$, it follows from (4.22) and $\mathbb{E}\{\|\tilde{z}_k\|^2\} > 0$ that

$$\mathbb{E}\{V(\eta_{k+1})|\eta_k\} - V(\eta_k) \leq -\kappa_3(\|\eta_k\|). \quad (4.25)$$

It follows directly from (4.16), (4.25), and Lemma 4.1 that the augmented system (4.6) is asymptotically stable in probability, and this concludes the proof.

Remark 4.2 *In Theorem 4.1, a nonlinear filter has been designed to guarantee the asymptotic stability in probability. The nonlinear function $\bar{\sigma}$ and the matrices $\tilde{\Upsilon}$ and $\tilde{\Upsilon}_i (i = 1, 2, 3)$ represent the influences of stochastic sensor saturations, and the parameter τ quantifies the effects of Markov time delays. The scalar r represents the consideration of the noise w_k. It is noted that the conditions established in (4.1) are in a very general form that will be applied to some special cases later.*

Lemma 4.2 *[66] Let $x, y \in \mathbb{R}^n$ and $\epsilon > 0$. Then we have*

$$2x^T y \leq \epsilon x^T x + \epsilon^{-1} y^T y. \quad (4.26)$$

In the case that $\bar{k}(\bar{x}) \equiv I$, the conditions of Theorem 4.1 can be further simplified/decoupled.

Corollary 4.1 *Given a disturbance attenuation level $\gamma > 0$ and $\bar{k}(\bar{x}) \equiv I$. If there exist two positive constants μ_1, μ_2, two positive definite matrices $P = P^T > 0$ and $Q = Q^T > 0$ and two $\mathcal{K}$ class functions κ_1 and κ_2 satisfying the*

Main Results

following inequalities for all $\tau \in N$:

$$C_\tau^T \hat{g}^T(\hat{x}, \tau) Q \hat{g}(\hat{x}, \tau) C_\tau \leq \mu_1 I, \tag{4.27}$$

$$\gamma^2 I - r^2 \bar{s}^T(\bar{x}) P \bar{s}(\bar{x}) - \bar{g}^T(\bar{x}) P \bar{g}(\bar{x}) > (\mu_1 + \mu_2) I, \tag{4.28}$$

$$\begin{aligned}\mathbb{H}_1(\bar{x}, \tau) =& \frac{5}{\mu_2} \left(\|\bar{f}^T(\bar{x}) P \bar{g}(\bar{x})\|^2 + r^4 \|\bar{h}^T(\bar{x}) P \bar{s}(\bar{x})\|^2 \right) + \left(\bar{f}(\bar{x}) + \bar{x} \right)^T P \\
& \times \left(\bar{f}(\bar{x}) - \bar{x} \right) + r^2 \bar{h}^T(\bar{x}) P \bar{h}(\bar{x}) + \left(\frac{5 \bar{v}^2 \mu_1^2}{\mu_2} + \bar{v} + 2 \bar{v} \mu_1 \right) \|\bar{\sigma}(\bar{l}(\bar{x}))\|^2 \\
& + \left(\frac{5(1-\bar{v})^2 \mu_1^2}{\mu_2} + (1-\bar{v}) + 2(1-\bar{v}) \mu_1 \right) \|\bar{l}(\bar{x})\|^2 \\
& + 2 \|\bar{m}(\bar{x})\|^2 + \kappa_1(\|\bar{x}\|) < 0,\end{aligned} \tag{4.29}$$

$$\begin{aligned}\mathbb{H}_2(\hat{x}, \tau) =& (\hat{f}(\hat{x}) + \hat{x})^T Q (\hat{f}(\hat{x}) - \hat{x}) + \left(\frac{5}{\mu_2} + 1 \right) \|\hat{f}^T(\hat{x}) Q \hat{g}(\hat{x}, \tau) C_\tau\|^2 \\
& + 2 \|\hat{m}(\hat{x})\|^2 + \kappa_2(\|\hat{x}\|) < 0,\end{aligned} \tag{4.30}$$

then the filtering problem for system (4.1) is solved by (4.5) in the sense of asymptotic stability in probability.

Proof With the element inequality $\|a + b\|^2 \leq 2\|a\|^2 + 2\|b\|^2$, we have

$$\|\tilde{z}\|^2 = \|\bar{m}(\bar{x}) - \hat{m}(\hat{x})\|^2 \leq 2\|\bar{m}(\bar{x})\|^2 + 2\|\hat{m}(\hat{x})\|^2. \tag{4.31}$$

Considering (4.17), (4.27), and (4.28), we have

$$\begin{aligned}A(\bar{x}, \hat{x}, \tau) =& \gamma^2 I - r^2 \bar{s}^T(\bar{x}) P \bar{s}(\bar{x}) - \bar{g}^T(\bar{x}) P \bar{g}(\bar{x}) - C_\tau^T \hat{g}^T(\hat{x}, \tau) Q \hat{g}(\hat{x}, \tau) C_\tau \\
\geq& (\mu_1 + \mu_2) I - \mu_1 I = \mu_2 I.\end{aligned} \tag{4.32}$$

Then, it follows from (4.18), (4.27), (4.28), and (4.32) that

$$\begin{aligned}B(\bar{x}, \hat{x}, \tau) A^{-1}(\bar{x}, \hat{x}, \tau) B^T(\bar{x}, \hat{x}, \tau) \leq & \frac{1}{\mu_2} B(\bar{x}, \hat{x}, \tau) B^T(\bar{x}, \hat{x}, \tau) \\
\leq & \frac{5}{\mu_2} \Big(\|\bar{f}^T(\bar{x}) P \bar{g}(\bar{x})\|^2 + r^4 \|\bar{h}^T(\bar{x}) P \bar{s}(\bar{x})\|^2 \\
& + \|\hat{f}^T(\hat{x}) Q \hat{g}(\hat{x}, \tau) C_\tau\|^2 + \bar{v}^2 \mu_1^2 \|\bar{\sigma}(\bar{l}(\bar{x}))\|^2 \\
& + (1 - \bar{v})^2 \mu_1^2 \|\bar{l}(\bar{x})\|^2 \Big).\end{aligned} \tag{4.33}$$

With Lemma 4.2, we can obtain

$$2\hat{f}^T(\hat{x}) Q \hat{g}(\hat{x}, \tau) C_\tau \bar{\Upsilon} \bar{\sigma}(\bar{l}(\bar{x})) \leq \bar{v} \left(\|\hat{f}^T(\hat{x}) Q \hat{g}(\hat{x}, \tau) C_\tau\|^2 + \|\bar{\sigma}(\bar{l}(\bar{x}))\|^2 \right), \tag{4.34}$$

$$2\hat{f}^T(\hat{x}) Q \hat{g}(\hat{x}, \tau) C_\tau (I - \bar{\Upsilon}) \bar{\sigma}(\bar{l}(\bar{x})) \leq (1 - \bar{v}) \left(\|\hat{f}^T(\hat{x}) Q \hat{g}(\hat{x}, \tau) C_\tau\|^2 + \|\bar{l}(\bar{x})\|^2 \right). \tag{4.35}$$

Define the $\mathcal{K}$ class function as $\kappa(\|\eta\|) = \kappa_1(\|\bar{x}\|) + \kappa_2(\|\hat{x}\|)$, and it follows that

$$\begin{aligned}\mathbb{H}(\bar{x},\hat{x},\tau) \leq & \frac{5}{\mu_2}\Big(\|\bar{f}^T(\bar{x})P\bar{g}(\bar{x})\|^2 + r^4\|\bar{h}^T(\bar{x})P\bar{s}(\bar{x})\|^2 + \|\hat{f}^T(\hat{x})Q\hat{g}(\hat{x},\tau)C_\tau\|^2 \\ & + \bar{v}^2\mu_1^2\|\bar{\sigma}(\bar{l}(\bar{x}))\|^2 + (1-\bar{v})^2\mu_1^2\|\bar{l}(\bar{x})\|^2\Big) + \big(\bar{f}(\bar{x})+\bar{x}\big)^T P \\ & \times \big(\bar{f}(\bar{x})-\bar{x}\big) + \big(\hat{f}(\hat{x})+\hat{x}\big)^T \times Q\big(\hat{f}(\hat{x})-\hat{x}\big) + r^2\bar{h}^T(\bar{x})P\bar{h}(\bar{x}) \\ & + \bar{v}\|\bar{\sigma}(\bar{l}(\bar{x}))\|^2 + \bar{v}\|\hat{f}^T(\hat{x})Q\hat{g}(\hat{x},\tau)C_\tau\|^2 + (1-\bar{v})\|\bar{l}(\bar{x})\|^2 \\ & + (1-\bar{v})\|\hat{f}^T(\hat{x})Q\hat{g}(\hat{x},\tau)C_\tau\|^2 + 2\bar{v}\mu_1\|\bar{\sigma}(\bar{l}(\bar{x}))\|^2 + 2(1-\bar{v})\mu_1 \\ & \times \|\bar{l}(\bar{x})\|^2 + 2\|\bar{m}(\bar{x})\|^2 + 2\|\hat{m}(\hat{x})\|^2 + \kappa_1(\|\bar{x}\|) + \kappa_2(\|\hat{x}\|) \\ = & \mathbb{H}_1(\bar{x},\tau) + \mathbb{H}_2(\hat{x},\tau) < 0.\end{aligned}$$

(4.36)

The rest of the proof follows directly from that of Theorem 4.1.

If the adopted positive definite function is dependent on not only the augmented states but also the values of time delays, then the sufficient conditions would be obtained subsequently.

Theorem 4.2 *Given a disturbance attenuation level $\gamma > 0$. If there exist two sets of positive-definite matrices $P(\tau) = P^T(\tau) > 0$ and $Q(\tau) = Q^T(\tau) > 0$ for all $\tau \in N$, and a $\mathcal{K}$ class function κ satisfying the inequalities*

$$\begin{cases}\mathbb{H}(\bar{x},\tilde{x},\tau) = B(\bar{x},\tilde{x},\tau)A^{-1}(\bar{x},\tilde{x},\tau)B^T(\bar{x},\tilde{x},\tau) + r^2\bar{h}^T(\bar{x})\tilde{P}(\tau)\bar{h}(\bar{x}) \\ \qquad + \bar{f}^T(\bar{x})\tilde{P}(\tau)\bar{f}(\bar{x}) + D(\bar{x},\tilde{x},\tau) - \bar{x}^T P(\tau)\bar{x} - \hat{x}^T Q(\tau)\hat{x} \\ \qquad + \|\tilde{z}\|^2 + \kappa(\|\eta\|) < 0, \text{for any } \bar{x},\hat{x} \in \mathbb{R}^{d\times n_x}, \tau \in N, \\ A(\bar{x},\tilde{x},\tau) > 0, \text{ for any } \bar{x},\hat{x} \in \mathbb{R}^{d\times n_x}, \tau \in N,\end{cases}$$

(4.37)

where $\tilde{P}(\tau) = \sum\limits_{i=0}^{d-1}\pi_{\tau i}P(i)$, $\tilde{Q}(\tau) = \sum\limits_{i=0}^{d-1}\pi_{\tau i}Q(i)$ *and*

$$A(\bar{x},\tilde{x},\tau) = \gamma^2 I - r^2\bar{s}^T(\bar{x})\tilde{P}(\tau)\bar{s}(\bar{x}) - \bar{g}^T(\bar{x})\tilde{P}(\tau)\bar{g}(\bar{x}) - \bar{k}^T(\bar{x})S_1(\hat{x},\tau)\bar{k}(\bar{x}),$$

(4.38)

$$\begin{aligned}B(\bar{x},\tilde{x},\tau) =& \bar{f}(\bar{x})\tilde{P}(\tau)\bar{g}(\bar{x}) + \hat{f}(\hat{x})\tilde{Q}(\tau)\hat{g}(\hat{x},\tau)C_\tau\bar{k}(\bar{x}) + r^2\bar{h}^T(\bar{x})\tilde{P}(\tau)\bar{s}(\bar{x}) \\ & + S_2^T(\bar{x},\hat{x},\tau)\tilde{Q}(\tau)\hat{g}(\hat{x},\tau)C_\tau\bar{k}(\bar{x}),\end{aligned}$$

(4.39)

$$\begin{aligned}D(\bar{x},\tilde{x},\tau) =& \hat{f}^T(\hat{x})\tilde{Q}(\tau)\hat{f}(\hat{x}) + 2\hat{f}^T(\hat{x})\tilde{Q}(\tau)S_2(\bar{x},\hat{x},\tau) \\ & + \bar{\sigma}^T(\bar{l}(\bar{x}))\big(\bar{\Upsilon}_1 \circ S_1(\hat{x},\tau)\big) \times \bar{\sigma}(\bar{l}(\bar{x})) \\ & + 2\bar{\sigma}^T(\bar{l}(\bar{x}))\big(\bar{\Upsilon}_3 \circ S_1(\hat{x},\tau)\big)\bar{l}(\bar{x}) + \bar{l}^T(\bar{x})\big(\bar{\Upsilon}_2 \circ S_1(\hat{x},\tau)\big)\bar{l}(\bar{x}),\end{aligned}$$

(4.40)

$$S_1(\hat{x},\tau) = C_\tau^T \hat{g}^T(\hat{x},\tau)\tilde{Q}(\tau)\hat{g}(\hat{x},\tau)C_\tau,$$

(4.41)

$$S_2(\bar{x},\hat{x},\tau) = \hat{g}(\hat{x},\tau)C_\tau\bar{\Upsilon}\bar{\sigma}(\bar{l}(\bar{x})) + \hat{g}(\hat{x},\tau)C_\tau(I - \bar{\Upsilon})\bar{l}(\bar{x}),$$

(4.42)

then the filtering problem for system (4.1) is solved by (4.5) in the sense of asymptotic stability in probability.

Proof *The positive definite function is taken as:*

$$V(\eta_k, d_k) = \bar{x}_k^T P(d_k)\bar{x}_k + \hat{x}_k^T Q(d_k)\hat{x}_k. \tag{4.43}$$

We take

$$\bar{\rho} = \max\left\{\max_{\tau \in N} \upsilon_{\max}(P(\tau)), \max_{\tau \in N} \upsilon_{\max}(Q(\tau))\right\}$$

and

$$\underline{\rho} = \min\left\{\min_{\tau \in N} \upsilon_{\min}(P(\tau)), \min_{\tau \in N} \upsilon_{\min}(Q(\tau))\right\},$$

then two $\mathcal{K}_\infty$ functions can be defined as $\kappa_1(\|\eta\|) = \underline{\rho}\|\eta\|^2$ and $\kappa_2(\|\eta\|) = \bar{\rho}\|\eta\|^2$, and it follows that $\kappa_1(\|\eta\|) \leq V(\eta, \tau) \leq \kappa_2(\|\eta\|)$ for all $\tau \in N$.

With notations above, one has

$$\mathbb{E}\{V(\eta_{k+1}, d_{k+1})|\eta_k, d_k\} - V(\eta_k, d_k) + \mathbb{E}\{\|\tilde{z}_k\|^2\} - \gamma^2 \mathbb{E}\{\|\bar{v}_k\|^2\} + \kappa(\|\eta_k\|)$$
$$= \mathbb{E}\Big\{-(\bar{v}_k - \bar{v}_k^*)^T A(\bar{x}_k, \tilde{x}_k, d_k)(\bar{v}_k - \bar{v}_k^*) + B(\bar{x}_k, \tilde{x}_k, d_k) A^{-1}(\bar{x}_k, \tilde{x}_k, d_k) B^T$$
$$\times (\bar{x}_k, \tilde{x}_k, d_k) + D(\bar{x}_k, \tilde{x}_k, d_k) + r^2 \bar{h}^T(\bar{x}_k)\tilde{P}(d_k)\bar{h}(\bar{x}_k) + \bar{f}^T(\bar{x}_k)\tilde{P}(d_k)\bar{f}(\bar{x}_k)$$
$$- \bar{x}_k^T P(d_k)\bar{x}_k - \hat{x}_k^T Q(d_k)\hat{x}_k + \|\tilde{z}_k\|^2 + \kappa(\|\eta_k\|)\Big\}$$
$$\leq \mathbb{H}(\bar{x}_k, \hat{x}_k, d_k) < 0.$$

The rest of the proof follows directly from that of Theorem 4.1 and is therefore omitted.

We also have the following corollary from Theorem 4.2 and Corollary 4.1.

Corollary 4.2 *Given a disturbance attenuation level $\gamma > 0$ and $\bar{k}(\bar{x}) \equiv I$. If there exist two positive constants μ_1, μ_2, two sets of positive definite matrices $P(\tau) = P^T(\tau) > 0$ and $Q(\tau) = Q^T(\tau) > 0$ for all $\tau \in N$, and two $\mathcal{K}$ class functions κ_1 and κ_2 satisfying the following inequalities for all $\tau \in N$:*

$$C_\tau^T \hat{g}^T(\hat{x}, \tau) \tilde{Q}(\tau) \hat{g}(\hat{x}, \tau) C_\tau \leq \mu_1 I, \tag{4.44}$$

$$\gamma^2 I - r^2 \bar{s}^T(\bar{x}) \tilde{P}(\tau) \bar{s}(\bar{x}) - \bar{g}^T(\bar{x}) \tilde{P}(\tau) \bar{g}(\bar{x}) > (\mu_1 + \mu_2) I, \tag{4.45}$$

$$\mathbb{H}_1(\bar{x}, \tau) = \frac{5}{\mu_2}\left(\|\bar{f}^T(\bar{x})\tilde{P}(\tau)\bar{g}(\bar{x})\|^2 + r^4\|\bar{h}^T(\bar{x})\tilde{P}(\tau)\bar{s}(\bar{x})\|^2\right) + \bar{f}^T(\bar{x})\tilde{P}(\tau)\bar{f}(\bar{x})$$
$$+ r^2 \bar{h}^T(\bar{x})\tilde{P}(\tau)\bar{h}(\bar{x}) - \bar{x}^T P(\tau)\bar{x} + \left(\frac{5\bar{v}^2 \mu_1^2}{\mu_2} + \bar{v} + 2\bar{v}\mu_1\right)\|\bar{\sigma}(\bar{l}(\bar{x}))\|^2$$
$$+ \left(\frac{5(1-\bar{v})^2 \mu_1^2}{\mu_2} + (1-\bar{v}) + 2(1-\bar{v})\mu_1\right)\|\bar{l}(\bar{x})\|^2$$
$$+ 2\|\bar{m}(\bar{x})\|^2 + \kappa_1(\|\bar{x}\|) < 0, \tag{4.46}$$

$$\mathbb{H}_2(\hat{x}, \tau) = \hat{f}^T(\hat{x})\tilde{Q}(\tau)\hat{f}(\hat{x}) - \hat{x}^T Q(\tau)\hat{x} + \left(\frac{5}{\mu_2} + 1\right)\|\hat{f}^T(\hat{x})\tilde{Q}(\tau)\hat{g}(\hat{x}, \tau)C_\tau\|^2$$
$$+ 2\|\hat{m}(\hat{x})\|^2 + \kappa_2(\|\hat{x}\|) < 0, \tag{4.47}$$

then the filtering problem for system (4.1) is solved by (4.5) in the sense of asymptotic stability in probability.

Proof The proof of the corollary is a straightforward consequence of that of Corollary 4.1 and is omitted here for conciseness.

Remark 4.3 *A set of nonlinear filters has been obtained via solving nonlinear matrix inequalities for some positive definite functions that could be either delay-dependent or delay-independent. Sufficient conditions have been achieved under which the estimation is asymptotically stable in probability in the disturbance-free case and robust to the exogenous disturbances under the zero-initial condition. In real-world applications, a linear filter is much easier to implement than a nonlinear one. As a result, a linear filter for the nonlinear system (4.1) would be investigated next.*

Consider the filter of the following structure:

$$\begin{cases} \hat{x}_{k+1} = F\hat{x}_k + G(d_k)y_k, \\ \hat{z}_k = M\hat{x}_k, \end{cases} \quad (4.48)$$

where $\hat{x}_k \in \mathbb{R}^{d \times n_x}$ is the state estimate; $\hat{z}_k \in \mathbb{R}^{n_z}$ is the estimate of z_k. F, G, and M are filter parameters of appropriate dimensions to be determined.

Similar to the nonlinear case, we can have the following augmented system.

$$\begin{cases} \eta_{k+1} = \tilde{f}(\eta_k, d_k) + \tilde{g}(\eta_k, d_k)v_k + (\tilde{h}(\eta_k) + \tilde{s}(\eta_k)\bar{v}_k)w_k, \\ \tilde{z}_k = \bar{m}(\bar{x}_k) - M\hat{x}_k, \end{cases} \quad (4.49)$$

where

$$\tilde{f}(\eta_k, d_k) = \begin{bmatrix} \bar{f}(\bar{x}_k) \\ F\hat{x}_k + G(d_k)C_{d_k}\Upsilon_k\bar{\sigma}(\bar{l}(\bar{x}_k)) + G(d_k)C_{d_k}(I - \Upsilon_k)\bar{l}(\bar{x}_k) \end{bmatrix},$$

$$\tilde{g}(\eta_k, d_k) = \begin{bmatrix} \bar{g}(\bar{x}_k) \\ G(d_k)C_{d_k}\bar{k}(\bar{x}_k) \end{bmatrix}, \tilde{h}(\eta_k) = \begin{bmatrix} \bar{h}(\bar{x}_k) \\ 0 \end{bmatrix}, \tilde{s}(\eta_k) = \begin{bmatrix} \bar{s}(\bar{x}_k) \\ 0 \end{bmatrix}.$$

In virtue of Theorem 4.1, the following results can be obtained.

Theorem 4.3 *Given a disturbance attenuation level $\gamma > 0$. If there exist two positive definite matrices $P = P^T > 0$ and $Q = Q^T > 0$, and a $\mathcal{K}$ class function κ_3 satisfying the inequalities*

$$\begin{cases} \mathbb{H}(\bar{x}, \hat{x}, \tau) = B(\bar{x}, \hat{x}, \tau)A^{-1}(\bar{x}, \hat{x}, \tau)B^T(\bar{x}, \hat{x}, \tau) + r^2\bar{h}^T(\bar{x})P\bar{h}(\bar{x}) \\ \qquad +D(\bar{x}, \hat{x}, \tau) + \|\tilde{z}\|^2 + \kappa_3(\|\eta\|) < 0, \text{ for any } \bar{x}, \hat{x} \in \mathbb{R}^{d \times n_x}, \tau \in N \\ A(\bar{x}, \tilde{x}, \tau) > 0, \text{ for any } \bar{x}, \tilde{x} \in \mathbb{R}^{d \times n_x}, \tau \in N, \end{cases}$$

$$(4.50)$$

Main Results

where

$$A(\bar{x}, \hat{x}, \tau) = \gamma^2 I - r^2 \bar{s}^T(\bar{x}) P \bar{s}(\bar{x}) - \bar{g}^T(\bar{x}) P \bar{g}(\bar{x}) - \bar{k}^T(\bar{x}) S_1(\tau) \bar{k}(\bar{x}), \quad (4.51)$$
$$B(\bar{x}, \hat{x}, \tau) = r^2 \bar{h}^T(\bar{x}) P \bar{s}(\bar{x}) + \bar{f}^T(\bar{x}) P \bar{g}(\bar{x}) + \hat{x}^T F^T Q G(\tau) C_\tau \bar{k}(\bar{x})$$
$$\qquad + S_2^T(\bar{x}, \tau) Q G(\tau) C_\tau \bar{k}(\bar{x}), \quad (4.52)$$
$$D(\bar{x}, \hat{x}, \tau) = (\bar{f}(\bar{x}) + \bar{x})^T P(\bar{f}(\bar{x}) - \bar{x}) + \hat{x}^T (F + I)^T Q(F - I)\hat{x}$$
$$\qquad + 2\hat{x}^T F^T Q S_2(\bar{x}, \tau) + \bar{\sigma}^T(\bar{l}(\bar{x}))\left(\tilde{\Upsilon}_1 \circ S_1(\tau)\right) \bar{\sigma}(\bar{l}(\bar{x}))$$
$$\qquad + 2\bar{\sigma}^T(\bar{l}(\bar{x}))\left(\tilde{\Upsilon}_3 \circ S_1(\tau)\right) \bar{l}(\bar{x}) + \bar{l}^T(\bar{x})\left(\tilde{\Upsilon}_2 \circ S_1(\tau)\right) \bar{l}(\bar{x}), \quad (4.53)$$
$$S_1(\tau) = C_\tau^T G^T(\tau) Q G(\tau) C_\tau, \quad (4.54)$$
$$S_2(\bar{x}, \tau) = G(\tau) C_\tau \tilde{\Upsilon} \bar{\sigma}(\bar{l}(\bar{x})) + G(\tau) C_\tau (I - \tilde{\Upsilon}) \bar{l}(\bar{x}), \quad (4.55)$$

then the filtering problem for system (4.1) is solved by (4.48) in the sense of asymptotic stability in probability.

Proof *This proof can be completed by following the similar line of Theorem 4.1 and is therefore omitted.*

Similar to Corollaries 4.1–4.2, we can have certain decoupled sufficient conditions in the following corollaries.

Corollary 4.3 *Given a disturbance attenuation level $\gamma > 0$ and $\bar{k}(\bar{x}) \equiv I$. If there exist two positive constants μ_1, μ_2, two positive definite matrices $P = P^T > 0$ and $Q = Q^T > 0$, and two $\mathcal{K}$ class functions κ_1 and κ_2 satisfying the following inequalities for all $\tau \in N$:*

$$C_\tau^T G^T Q G C_\tau \leq \mu_1 I, \quad (4.56)$$
$$\gamma^2 I - r^2 \bar{s}^T(\bar{x}) P \bar{s}(\bar{x}) - \bar{g}^T(\bar{x}) P \bar{g}(\bar{x}) > (\mu_1 + \mu_2) I, \quad (4.57)$$
$$\mathbb{H}_1(\bar{x}, \tau) = \frac{5}{\mu_2}\left(\|\bar{f}^T(\bar{x}) P \bar{g}(\bar{x})\|^2 + r^4 \|\bar{h}^T(\bar{x}) P \bar{s}(\bar{x})\|^2\right) + \left(\bar{f}(\bar{x}) + \bar{x}\right)^T P$$
$$\qquad \times \left(\bar{f}(\bar{x}) - \bar{x}\right) + r^2 \bar{h}^T(\bar{x}) P \bar{h}(\bar{x}) + \left(\frac{5\bar{v}^2 \mu_1^2}{\mu_2} + \bar{v} + 2\bar{v}\mu_1\right) \|\bar{\sigma}(\bar{l}(\bar{x}))\|^2$$
$$\qquad + \left(\frac{5(1-\bar{v})^2 \mu_1^2}{\mu_2} + (1-\bar{v}) + 2(1-\bar{v})\mu_1\right) \|\bar{l}(\bar{x})\|^2$$
$$\qquad + 2\|\bar{m}(\bar{x})\|^2 + \kappa_1(\|\bar{x}\|) < 0, \quad (4.58)$$

$$\mathbb{H}_2(\hat{x}, \tau) = \hat{x}^T (F + I)^T Q (F - I) \hat{x} + \left(\frac{5}{\mu_2} + 1\right) \|\hat{x}^T F^T Q G C_\tau\|^2 + 2\|M\hat{x}\|^2$$
$$\qquad + \kappa_2(\|\hat{x}\|) < 0, \quad (4.59)$$

then the filtering problem for system (4.1) is solved by (4.48) in the sense of asymptotic stability in probability where $G(\tau) = G$ for any $\tau \in N$.

Corollary 4.4 *Given a disturbance attenuation level $\gamma > 0$ and $\bar{k}(\bar{x}) \equiv I$. If there exist two positive constants μ_1, μ_2, two sets of positive-definite matrices $P(\tau) = P^T(\tau) > 0$ and $Q(\tau) = Q^T(\tau) > 0$ for all $\tau \in N$, and two $\mathcal{K}$ class functions κ_1 and κ_2 satisfying the following inequalities for all $\tau \in N$:*

$$C_\tau^T G^T(\tau) \tilde{Q}(\tau) G(\tau) C_\tau \leq \mu_1 I, \tag{4.60}$$

$$\gamma^2 I - r^2 \bar{s}^T(\bar{x}) \tilde{P}(\tau) \bar{s}(\bar{x}) - \bar{g}^T(\bar{x}) \tilde{P}(\tau) \bar{g}(\bar{x}) > (\mu_1 + \mu_2) I, \tag{4.61}$$

$$\mathbb{H}_1(\bar{x}, \tau) = \frac{5}{\mu_2} \left(\|\bar{f}^T(\bar{x}) \tilde{P}(\tau) \bar{g}(\bar{x})\|^2 + r^4 \|\bar{h}^T(\bar{x}) \tilde{P}(\tau) \bar{s}(\bar{x})\|^2 \right) + \bar{f}^T(\bar{x}) \tilde{P}(\tau) \bar{f}(\bar{x})$$
$$+ r^2 \bar{h}^T(\bar{x}) \tilde{P}(\tau) \bar{h}(\bar{x}) - \bar{x}^T P(\tau) \bar{x} + \left(\frac{5\bar{v}^2 \mu_1^2}{\mu_2} + \bar{v} + 2\bar{v}\mu_1 \right) \|\bar{\sigma}(\bar{l}(\bar{x}))\|^2$$
$$+ \left(\frac{5(1-\bar{v})^2 \mu_1^2}{\mu_2} + (1-\bar{v}) + 2(1-\bar{v})\mu_1 \right) \|\bar{l}(\bar{x})\|^2$$
$$+ 2\|\bar{m}(\bar{x})\|^2 + \kappa_1(\|\bar{x}\|) < 0, \tag{4.62}$$

$$\mathbb{H}_2(\hat{x}, \tau) = \hat{x}^T F^T \tilde{Q}(\tau) F \hat{x} - \hat{x}^T Q(\tau) \hat{x} + \left(\frac{5}{\mu_2} + 1 \right) \|\hat{x}^T F^T \tilde{Q}(\tau) \hat{g}(\hat{x}, \tau) C_\tau\|^2$$
$$+ 2\|M\hat{x}\|^2 + \kappa_2(\|\hat{x}\|) < 0, \tag{4.63}$$

then the filtering problem for system (4.1) is solved by (4.48) in the sense of asymptotic stability in probability.

The proofs of Corollaries 4.3 and 4.4 follow directly from those of Corollaries 4.1 and 4.2, and are therefore omitted.

Remark 4.4 *The filtering problem for a class of nonlinear systems with stochastic sensor saturations and Markov time delays has been investigated in the sense of asymptotic stability in probability. Sufficient conditions have been established to guarantee the asymptotic stability in probability of the estimation process in the noise-free case and the robustness of the filtering error to the exogenous disturbances under the zero-initial condition. Specifically, the linear filters have been considered for the convenience of practical applications. Both time-dependent and time-independent linear filters have been obtained by choosing proper positive definite functions.*

4.3 Simulation Examples

4.3.1 Example A

Consider the following nonlinear system

$$\begin{cases} x_{k+1} = 0.7 x_k + 0.2 \sin(x_k) + v_k + 0.4 \cos(x_k) v_k w_k, \\ z_k = \frac{2}{3} x_k, \\ \tilde{y}_k = v_k \sigma(l(x_k)) + (1 - v_k) l(x_k) + v_k, \\ y_k = \tilde{y}_{k - d_k}, \end{cases} \tag{4.64}$$

Simulation Examples

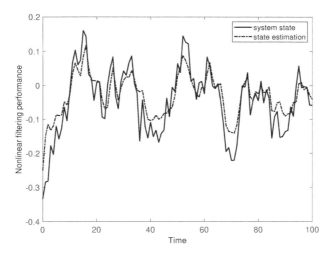

FIGURE 4.1: Nonlinear filtering performance

where d_k belongs to the set $\{0,1\}$ and its transition probability matrix is given by $\Pi = [\text{col}\{0.7, 0.8\}, \text{col}\{0.3, 0.2\}]$. The variance of w_k is 1 and $\text{Prob}\{v_k = 1\} = 0.9$. The saturation level is 0.1. The prescribed disturbance attenuation level is set to be $\gamma = 1.414$. Choose $\kappa_1(\|\bar{x}\|) = 0.01\|\bar{x}\|^2$, $\kappa_2(\|\hat{x}\|) = 0.01\|\hat{x}\|^2$, $\mu_1 = 0.5$, and $\mu_2 = 1$. Then, we can get the following nonlinear filter with Corollary 4.1:

$$\begin{cases} \hat{x}_{k+1} = 0.4\hat{x}_k \sin(\hat{x}_{k-1}) + 0.65 y_k, \\ \hat{z}_k = 0.5\hat{x}_k + 0.3\hat{x}_{k-1}. \end{cases} \quad (4.65)$$

As shown in Fig. 4.1, the established nonlinear filter can estimate z_k well.

4.3.2 Example B

An inverted pendulum example is presented in this subsection to demonstrate the effectiveness of the proposed approach. An appropriate controller has been predesigned to stabilize the system. The model of the inverted pendulum system is given by [94]

$$ml^2\ddot{\theta} - mgl\sin(\theta) + (\varsigma + \omega)\dot{\theta} + \kappa\theta = u + 2\nu_2, \quad (4.66)$$

where m is the mass, l is the length of the inverted pendulum, g is the gravitation coefficient, θ is the inclination angle, ς is the spring coefficient, κ is the damping parameter, ω is the white noise for the damping coefficient, ν_2 is the external disturbance, and u is the control input that has been predesigned as

$u = k_1\theta + k_2 ml^2\dot{\theta}$. Two output measurements without stochastic sensor saturation or transmission delays are $\tilde{y}_1 = \theta + \nu_1$ and $\tilde{y}_2 = ml^2\dot{\theta} + \dfrac{\sin(ml^2\dot{\theta})}{ml^2 g} + \nu_2$, respectively. The regulated output is described by $z = \dfrac{\theta + ml^2\dot{\theta}}{10}$. Since inertial sensors in reality usually undergo saturations [28], it is reasonable to consider sensor saturations in the inverted pendulum example (4.66) whose outputs are related to angle and angular acceleration.

Taking $x_1 = \theta$, $x_2 = ml^2\dot{\theta}$, and the sampling period as T, the system model can be discretized and realized by the state-space model as follows:

$$\begin{bmatrix} x_{1,k+1} \\ x_{2,k+1} \end{bmatrix} = \begin{bmatrix} 1 & \dfrac{T}{ml^2} \\ -\kappa T + Tk_1 & 1 - \dfrac{T\varsigma}{ml^2} + Tk_2 \end{bmatrix} \begin{bmatrix} x_{1,k} \\ x_{2,k} \end{bmatrix} + \begin{bmatrix} 0 \\ Tmgl\sin(x_{1,k}) \end{bmatrix}$$

$$+ \begin{bmatrix} 0 & 0 \\ 0 & 2T \end{bmatrix} \begin{bmatrix} \nu_{1,k} \\ \nu_{2,k} \end{bmatrix} + \begin{bmatrix} 0 & 0 \\ 0 & -\dfrac{\sqrt{T}}{ml^2} \end{bmatrix} \begin{bmatrix} x_{1,k} \\ x_{2,k} \end{bmatrix} \omega_k. \qquad (4.67)$$

The output measurement equation with stochastic sensor saturation is discretized as

$$\begin{bmatrix} \tilde{y}_{1,k} \\ \tilde{y}_{2,k} \end{bmatrix} = v_k \sigma \left(\begin{bmatrix} 1 & 0 \\ 0 & 1 \end{bmatrix} \begin{bmatrix} x_{1,k} \\ x_{2,k} \end{bmatrix} + \begin{bmatrix} 0 \\ \dfrac{\sin(x_{2,k})}{ml^2 g} \end{bmatrix} \right) + (1 - v_k) \left(\begin{bmatrix} 1 & 0 \\ 0 & 1 \end{bmatrix} \begin{bmatrix} x_{1,k} \\ x_{2,k} \end{bmatrix} \right.$$

$$\left. + \begin{bmatrix} 0 \\ \dfrac{\sin(x_{2,k})}{ml^2 g} \end{bmatrix} \right) + \begin{bmatrix} 1 & 0 \\ 0 & 1 \end{bmatrix} \begin{bmatrix} \nu_{1,k} \\ \nu_{2,k} \end{bmatrix}. \qquad (4.68)$$

Due to the delay, the received measurement is $y_k = \tilde{y}_{k-d_k}$, where d_k is the random time delay governed by a Markov chain. The state d_k belongs to the set $\{0, 1, 2\}$ and its transition probability matrix is given by $\Pi = [\text{col}\{0.2, 0.1, 0.2\}, \text{col}\{0.8, 0.4, 0.4\}, \text{col}\{0, 0.5, 0.4\}]$. Furthermore, the regulated output can be discretized as $z_k = 0.1x_{1,k} + 0.1x_{2,k}$. The system parameters are $m = 0.5$kg, $l = 0.5$m, $\varsigma = 0.25$, $k_1 = -49.5$, $k_2 = -167.5$, sampling period $T = 0.01$s, and $\kappa = 0.5$N/m. The variance of ω is 0.01 and $\text{Prob}\{v_k = 1\} = 0.2$. The saturation level is 1. The prescribed disturbance attenuation level is set to be $\gamma = 0.707$.

Consider the linear filter in the form of (4.48) and let $\kappa_1(\|\bar{x}\|) = 0.5\|\bar{x}\|^2$, $\kappa_2(\|\hat{x}\|) = 0.5\|\hat{x}\|^2$. F and M have been selected to reflect the linear part of the system dynamics as follows:

$$F = \begin{bmatrix} F_0 & 0 & 0 \\ I & 0 & 0 \\ 0 & I & 0 \end{bmatrix}, \quad M = \begin{bmatrix} M_0 & 0 & 0 \end{bmatrix},$$

Simulation Examples

where

$$F_0 = \begin{bmatrix} 1 & \dfrac{T}{ml^2} \\ -\kappa T + Tk_1 & 1 - \dfrac{T\varsigma}{ml^2} + Tk_2 \end{bmatrix}, \quad M_0 = \begin{bmatrix} 0.1 & 0.1 \end{bmatrix}.$$

In such a case, G can be calculated using Matlab for both delay-dependent and delay-independent filters. Using Corollaries 4.3 and 4.4 with $\mu_1 = \mu_2 = 0.15$, the feasible solutions for G can be obtained as

$$G = \begin{bmatrix} -0.0021 & -0.0031 & 0.0018 & -0.0034 & -0.0032 & 0.0009 \\ 0.0007 & -0.0119 & -0.0045 & -0.0125 & 0.0024 & 0.0003 \end{bmatrix}$$

for the delay-independent filter, and

$$G(0) = \begin{bmatrix} -0.0032 & -0.0013 & -0.0013 & -0.0014 & -0.0033 & 0.0003 \\ 0.0002 & -0.0077 & -0.0003 & -0.0083 & 0.0011 & -0.0048 \end{bmatrix},$$

$$G(1) = \begin{bmatrix} -0.0012 & -0.0028 & 0.0042 & -0.0028 & -0.0017 & 0.0017 \\ 0.0004 & -0.0079 & -0.0010 & -0.0097 & 0.0032 & 0.0003 \end{bmatrix},$$

$$G(2) = \begin{bmatrix} -0.0016 & -0.0024 & 0.0022 & -0.0021 & -0.0013 & 0.0007 \\ 0.0003 & -0.0076 & -0.0006 & -0.0094 & 0.0019 & -0.0012 \end{bmatrix}$$

for the delay-dependent filter.

As illustrated in Fig. 4.2, both the delay-dependent and delay-independent filters achieve acceptable estimation performance. The filtering errors are

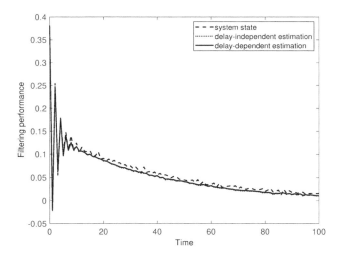

FIGURE 4.2: Delay-dependent and delay-independent filtering performances

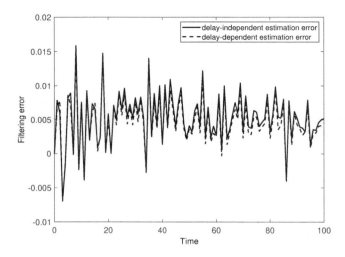

FIGURE 4.3: Delay-dependent and delay-independent filtering errors

depicted in Fig. 4.3. The average filtering error obtained with the delay-dependent filter is 5.258×10^{-3} in 100 Monte-Carlo simulations. With the delay-independent filter, the average filtering error is 5.789×10^{-3}. It can be easily seen that the delay dependent filter can generate smaller filtering error compared with the delay-independent one due to the consideration of the information on time delays.

4.4 Conclusion

In this chapter, the filtering problem has been investigated for a class of nonlinear systems with stochastic sensor saturations and Markovian measurement transmission delays. An H_∞ requirement has been considered in the sense of asymptotic stability in probability. A Bernoulli distributed sequence and a discrete-time Markov chain with finite states have been introduced to govern the random sensor saturations and the transmission time delays, respectively. Sufficient nonlinear conditions have been achieved to guarantee that the filtering process is asymptotically stable in probability in the disturbance-free case and satisfies the H_∞ criterion with respect to nonzero exogenous disturbances under the zero-initial condition. The results have been specialized to delay-independent and delay-dependent linear filters as well. Two simulation examples have been presented to show the effectiveness of the proposed algorithms.

5

Observer Design for Systems with Unknown Inputs and Missing Measurements

The observer design problem for systems with unknown inputs has been attracting considerable research attention in the past decades. Unknown inputs may arise from exogenous disturbances and/or unmodeled uncertainties in real-world systems and degrade the estimation performance severely. Now there has been a rich literature on the control/filtering problems for systems with unknown inputs such as multi-agent systems, time-delayed systems, and descriptor systems. There have been mainly two ways of coping with unknown inputs, which are reconstructing unknown inputs and eliminating the effects of unknown inputs. The unknown input observers which can completely decouple the estimation error from unknown inputs have been widely used to solve the fault detection and isolation problems as well. It is noted that unknown inputs have mostly been taken into account in the deterministic framework.

In many practical applications, missing measurement phenomenon occurs frequently due to accidental sensor failures or network communication delay/loss, etc. The phenomenon is also known as intermittent observation or data dropout. A classical approach to formulate the probabilistic missing measurements has been utilizing a Bernoulli sequence taking values on 0 or 1. The estimation problem for systems with missing measurements has been extensively studied [163], and the influences of the missing measurements phenomenon on the estimation performance have been discussed in detail [83]. Unfortunately, the state estimation problem for systems with both unknown inputs and missing measurements has not been fully investigated, not to mention the performance analysis with respect to boundedness.

It should be pointed out that, when missing measurements are considered, the estimation error cannot be completely decoupled from the unknown inputs as in deterministic systems. The randomness introduced by the missing measurements may result in that the impacts from the unknown disturbances on the estimation error are stochastic. As such, when only some statistical information on the missing measurements is available, the effects of unknown inputs cannot be fully eliminated from the estimation error. In such a case, it makes more sense to study the decoupling relationship in the mathematical expectation sense. Furthermore, the missing measurements will pose extra

DOI: 10.1201/9781003309482-5

challenges to the quantitative boundedness analysis. These difficulties constitute the main motivation of this chapter.

In this chapter, the observer design problem is addressed for a class of systems with unknown inputs and missing measurements. A Bernoulli distributed sequence that takes values on 0 or 1 is employed to govern the stochastic missing measurements. The observer is attained that can decouple the estimation error from the unknown disturbances in the mathematical expectation sense. Then, the uniform ultimate boundedness of the estimation error is investigated in the presence of unknown inputs and missing measurements. Finally, a simulation example is provided to show the effectiveness of the proposed method.

The main contributions of the chapter are summarized as follows: 1) the system model is comprehensive that covers unknown inputs and missing measurements; 2) the observer parameters are calculated to decouple the the mathematical expectations of estimation error from the unknown disturbances; and 3) the uniform ultimate boundedness of the estimation error is studied.

5.1 Problem Formulation

Consider the following discrete-time system:

$$\begin{cases} x_{k+1} = Ax_k + Bu_k + Ed_k + w_k, \\ y_k = \alpha_k Cx_k + v_k, \end{cases} \quad (5.1)$$

where $x_k \in \mathbb{R}^n$ is the state; $u_k \in \mathbb{R}^p$ is the control input; $y_k \in \mathbb{R}^m$ is measurement output; $d_k \in \mathbb{R}^q$ is the unknown external disturbance; $w_k \in \mathbb{R}^n$ and $v_k \in \mathbb{R}^m$ are the process noise and transmission noise, respectively; A, B, C, and E are known matrices with appropriate dimensions.

$\alpha_k \in \mathbb{R}$ is a Bernoulli distributed white sequence taking values on 0 or 1 with

$$\begin{cases} \text{Prob}\{\alpha_k = 1\} = \lambda, \\ \text{Prob}\{\alpha_k = 0\} = 1 - \lambda, \end{cases} \quad (5.2)$$

where $\lambda \in (0, 1]$ is a known scalar. The sequence α_k is introduced to govern the stochastic missing measurements. When $\alpha_k = 1$, the measurement with useful information can be utilized; when $\alpha_k = 0$, the received signal contains transmission noise only.

For system (5.1), the following observer is considered as a discrete version of that in [23]:

$$\begin{cases} z_{k+1} = Fz_k + TBu_k + Ky_k, \\ \hat{x}_{k+1} = z_{k+1} + Hy_{k+1}, \end{cases} \quad (5.3)$$

Observer Design 81

where $z_k \in \mathbb{R}^n$ is the observer state; $\hat{x}_k \in \mathbb{R}^n$ is the state estimate. Denote $e_k = x_k - \hat{x}_k$ as the estimation error.

Owing to the randomness brought in by the data dropout, the estimation error cannot be completely decoupled from d_k in the presence of the stochastic variable α_k no matter how the observer parameters are determined. Alternatively, the influences of d_k can be eliminated from the estimation error in the mathematical expectation sense by appropriately designing the observer, where some statistical information of the missing measurements will be utilized. In the next section, an observer in the form of (5.3) for system (5.1) will be developed such that the mathematical expectation of the estimation error is not subject to d_k.

5.2 Observer Design

In this section, the observer (5.3) will be established for system (5.1) with missing measurements.

Theorem 5.1 *Consider the observer (5.3) for system (5.1) with unknown inputs and missing measurements. If the observer parameters are selected as follows:*

$$K = K_1 + K_2, \tag{5.4}$$
$$E = \lambda HCE, \tag{5.5}$$
$$T = I - \lambda HC, \tag{5.6}$$
$$F = A - \lambda HCA - \lambda K_1 C, \tag{5.7}$$
$$K_2 = FH, \tag{5.8}$$

and all the eigenvalues of the system matrix F are stable, then the mathematical expectation of the estimation error will be completely decoupled from unknown inputs.

Proof Based on (5.1) and (5.3), we have

$$e_{k+1} = x_{k+1} - (z_{k+1} + Hy_{k+1}). \tag{5.9}$$

It follows that

$$\begin{aligned} e_{k+1} =& x_{k+1} - (\alpha_{k+1} HC x_{k+1} + H v_{k+1}) \\ & - [F z_k + T B u_k + (K_1 + K_2) y_k]. \end{aligned} \tag{5.10}$$

Then, we have

$$\begin{aligned} e_{k+1} =& (I - \alpha_{k+1} HC) x_{k+1} - H v_{k+1} - F(x_k - e_k) \\ & - H y_k) - T B u_k - K_1(\alpha_k C x_k + v_k) - K_2 y_k. \end{aligned} \tag{5.11}$$

According to (5.1), (5.11) can be written as

$$\begin{aligned}e_{k+1} =& Fe_k - Hv_{k+1} - K_1v_k + (I - \alpha_{k+1}HC)w_k - [F - (I - \alpha_{k+1}HC)A \\&+ \alpha_k K_1 C]x_k + [T - (I - \alpha_{k+1}HC)]Bu_k + (FH - K_2)y_k \\&+ (I - \alpha_{k+1}HC)Ed_k.\end{aligned} \quad (5.12)$$

Considering (5.8) and taking mathematical expectation of both sides of (5.12), we can get

$$\begin{aligned}\mathbb{E}\{e_{k+1}\} =& F\mathbb{E}\{e_k\} - Hv_{k+1} - K_1v_k + (I - \lambda HC)w_k - [F - (I - \lambda HC)A \\&+ \lambda K_1 C]x_k + [T - (I - \lambda HC)]Bu_k + (I - \lambda HC)Ed_k.\end{aligned} \quad (5.13)$$

From (5.5)–(5.7), it follows that

$$\mathbb{E}\{e_{k+1}\} = F\mathbb{E}\{e_k\} - Hv_{k+1} - K_1v_k + (I - \lambda HC)w_k, \quad (5.14)$$

which means that the mathematical expectation of the estimation error is fully decoupled from the unknown inputs. Moreover, since F is stable, the estimation error will be asymptotically stable in the absence of the noises and missing measurements phenomenon. This concludes the proof.

It is also noted that the necessary and sufficient conditions for the existence of the solutions to (5.4)–(5.8) are

- rank(CE)=rank(E).

- The pair $(C, A - \lambda HCA)$ is detectable.

The conditions are similar with those proposed in [23]. With the observer obtained in Theorem 1, the boundedness of the estimation error will be discussed in the next section.

5.3 Boundedness Analysis

Before proceeding further, the following definition is employed as a discrete version of that in [118, 125].

Definition 1: For nonlinear system $\beta_{k+1} = h(\beta_k)$, its solution is said to be uniformly ultimately bounded, if there exists a compact set $\Omega \in \mathbb{R}^n$ such that for all $\beta_0 \in \Omega$, there exist a bound Λ and a time $T(\Lambda, \beta_0)$ such that $\|\beta_k - \bar{\beta}\| \leq \Lambda$ for all $k \geq T$, where $\bar{\beta}$ is an equilibrium point.

With the definition above, the next theorem can be achieved.

Theorem 5.2 *Consider system (5.1) and the observer established in Theorem 1. If the disturbances w_k, v_k, d_k, system state x_k, control input u_k, and all the parameters (including the system parameters and observer parameters) are uniformly norm-bounded, then the estimation error e_k is uniformly ultimately bounded.*

Boundedness Analysis 83

Proof *Choose the following Lyapunov function*

$$V\{e_k\} = e_k^T e_k. \tag{5.15}$$

Then, we have

$$\begin{aligned}
&V\{e_{k+1}\} - V\{e_k\} \\
&= \Big\{ Fe_k - Hv_{k+1} - K_1 v_k + (I - \alpha_{k+1} HC) w_k - \big[F - (I - \alpha_{k+1} HC) A \\
&\quad + \alpha_k K_1 C \big] x_k + \big[T - (I - \alpha_{k+1} HC) \big] Bu_k \Big\}^T \Big\{ Fe_k - Hv_{k+1} - K_1 v_k \\
&\quad + (I - \alpha_{k+1} HC) w_k - [F - (I - \alpha_{k+1} HC) A + \alpha_k K_1 C] x_k \\
&\quad + [T - (I - \alpha_{k+1} HC)] Bu_k \Big\} - e_k^T e_k. \tag{5.16}
\end{aligned}$$

Based on (5.4)–(5.8), (5.16) can be written as

$$\begin{aligned}
&V\{e_{k+1}\} - V\{e_k\} \\
&= \Big\{ Fe_k - Hv_{k+1} - K_1 v_k + (I - \alpha_{k+1} HC) w_k - [(\alpha_{k+1} - \lambda) HCA \\
&\quad + (\alpha_k - \lambda) K_1 C] x_k + (\alpha_{k+1} - \lambda) HCBu_k \Big\}^T \Big\{ Fe_k - Hv_{k+1} - K_1 v_k \\
&\quad + (I - \alpha_{k+1} HC) w_k - [(\alpha_{k+1} - \lambda) HCA + (\alpha_k - \lambda) K_1 C] x_k \\
&\quad + (\alpha_{k+1} - \lambda) HCBu_k \Big\} - e_k^T e_k. \tag{5.17}
\end{aligned}$$

Since all the eigenvalues of F are strictly stable, there always exist positive scalars $\varepsilon, \gamma > 0$ such that $(1+\varepsilon) F^T F - I < -\gamma I$, and it follows that

$$\begin{aligned}
&V\{e_{k+1}\} - V\{e_k\} \\
&\leq e_k^T [(1+\varepsilon) F^T F - I] e_k + (1+\varepsilon^{-1}) \Big\{ -Hv_{k+1} - K_1 v_k + (I - \alpha_{k+1} HC) w_k \\
&\quad - [(\alpha_{k+1} - \lambda) HCA + (\alpha_k - \lambda) K_1 C] x_k + (\alpha_{k+1} - \lambda) HCBu_k \Big\}^T \Big\{ -Hv_{k+1} \\
&\quad - K_1 v_k + (I - \alpha_{k+1} HC) w_k - [(\alpha_{k+1} - \lambda) HCA + (\alpha_k - \lambda) K_1 C] x_k \\
&\quad + (\alpha_{k+1} - \lambda) HCBu_k \Big\}. \tag{5.18}
\end{aligned}$$

Applying element inequality to (5.18) yields that

$$\begin{aligned}
&V\{e_{k+1}\} - V\{e_k\} \\
&\leq e_k^T [(1+\varepsilon) F^T F - I] e_k + 5(1+\varepsilon^{-1}) v_{k+1}^T H^T H v_{k+1} + 5(1+\varepsilon^{-1}) v_k^T K_1^T K_1 v_k \\
&\quad + 5(1+\varepsilon^{-1}) w_k^T (I - \alpha_{k+1} HC)^T (I - \alpha_{k+1} HC) w_k + 5(1+\varepsilon^{-1}) x_k^T [(\alpha_{k+1} \\
&\quad - \lambda) HCA + (\alpha_k - \lambda) K_1 C]^T [(\alpha_{k+1} - \lambda) HCA + (\alpha_k - \lambda) K_1 C] x_k \\
&\quad + 5(1+\varepsilon^{-1}) (\alpha_{k+1} - \lambda)^2 u_k^T B^T C^T H^T HCBu_k. \tag{5.19}
\end{aligned}$$

It follows that

$$V\{e_{k+1}\} - V\{e_k\}$$
$$\leq e_k^T[(1+\varepsilon)F^TF - I]e_k + 5(1+\varepsilon^{-1})v_{k+1}^T H^T H v_{k+1} + 5(1+\varepsilon^{-1})v_k^T K_1^T K_1 v_k$$
$$+ 5(1+\varepsilon^{-1})w_k^T(I - \alpha_{k+1}HC)^T(I - \alpha_{k+1}HC)w_k + 10(1+\varepsilon^{-1})(\alpha_{k+1}$$
$$- \lambda)^2 x_k^T A^T C^T H^T HCA x_k + 10(1+\varepsilon^{-1})(\alpha_k - \lambda)^2 x_k^T C^T K_1^T K_1 C x_k$$
$$+ 5(1+\varepsilon^{-1})(\alpha_{k+1} - \lambda)^2 u_k^T B^T C^T H^T HCB u_k. \tag{5.20}$$

According to the assumptions made in Theorem 2, we can denote

$$\|w_k\| \leq \rho_w, \|v_k\| \leq \rho_v, \|d_k\| \leq \rho_d, \|x_k\| \leq \rho_x, \|u_k\| \leq \rho_u,$$
$$\varrho_1 = \max\{\|I - HC\|, 1\}, \bar{\lambda} = \max\{\lambda, 1-\lambda\}, \varrho_2 = \|HCA\|, \varrho_3 = \|K_1C\|,$$
$$\varrho_4 = \|HCB\|, \varrho_5 = \|K_1\|, \varrho_6 = \|H\|. \tag{5.21}$$

Then, we can get

$$V\{e_{k+1}\} - V\{e_k\}$$
$$\leq -\gamma e_k^T e_k + 5(1+\varepsilon^{-1})\rho_v^2(\varrho_5^2 + \varrho_6^2) + 5(1+\varepsilon^{-1})\rho_w^2 \varrho_1^2$$
$$+ 10(1+\varepsilon^{-1})\rho_x^2(\varrho_2^2 + \varrho_3^2) + 5(1+\varepsilon^{-1})\bar{\lambda}^2 \rho_u^2 \varrho_4^2. \tag{5.22}$$

Let

$$\sigma = 5(1+\varepsilon^{-1})\rho_v^2(\varrho_5^2 + \varrho_6^2) + 5(1+\varepsilon^{-1})\rho_w^2 \varrho_1^2$$
$$+ 10(1+\varepsilon^{-1})\rho_x^2(\varrho_2^2 + \varrho_3^2) + 5(1+\varepsilon^{-1})\bar{\lambda}^2 \rho_u^2 \varrho_4^2. \tag{5.23}$$

Provided that

$$\|e_k\| > \sqrt{\frac{\sigma}{\gamma}}, \tag{5.24}$$

then $V\{e_{k+1}\} - V\{e_k\} < 0$. Using the standard Lyapunov extension theorem, we can assert that the estimation error e_k is uniformly ultimately bounded, and this completes the proof.

So far, the observer design problem has been solved for a class of systems with unknown inputs and missing measurements. Because of the randomness introduced by the missing measurements, the estimation error cannot be fully decoupled from the unknown inputs. The observer has been derived which can eliminate the effects of unknown inputs in the mathematical expectation sense. The scalar λ quantifies the missing measurement phenomenon in the observer parameters. Furthermore, the uniform ultimate boundedness of the estimation error has been analyzed rigorously, and the relationship between the bound and the system parameters has also been reflected in Theorem 2. It is worth mentioning that the filter design algorithm and the performance analysis can be readily extended to time-varying systems. In the next section, a simulation example will be presented to show the effectiveness of the proposed approach.

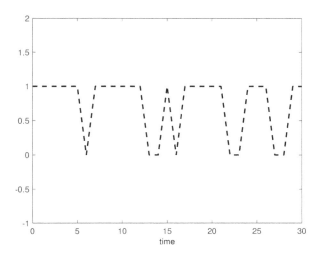

FIGURE 5.1: The values of α_k

5.4 Illustrative Examples

Consider system (5.1) with the following parameters:

$$A = \begin{bmatrix} 0.97 & -0.04 \\ 0 & -0.72 \end{bmatrix}, B = \begin{bmatrix} 1 \\ 0 \end{bmatrix}, E = \begin{bmatrix} 1 \\ 1 \end{bmatrix}, C = \begin{bmatrix} 1 & 0 \end{bmatrix}.$$

The closed-loop control is designed as $u_k = -0.2y_k$. The disturbances d_k, w_k, and v_k are mutually independent Gaussian distributed sequences whose variances are all $1 \times 10^{-12} I$. λ is selected as 0.85, and the initial state is set to be $[0.1, 0.1]^T$. Then the observer parameters in (5.4)–(5.8) are determined as

$$H = \begin{bmatrix} 1.1765 \\ 1.1765 \end{bmatrix}, K_1 = \begin{bmatrix} 0.4 \\ 0.3 \end{bmatrix}, K_2 = \begin{bmatrix} -0.4 \\ -2.24 \end{bmatrix},$$

$$T = \begin{bmatrix} 0 & 0 \\ -1 & 1 \end{bmatrix}, F = \begin{bmatrix} -0.34 & 0 \\ -1.225 & -0.68 \end{bmatrix}.$$

In such a case, the eigenvalues of F are obviously stable.

Fig. 5.1 plots the values of α_k. Figs. 5.2–5.3 depict the states and their estimates obtained with Theorem 1. It can be seen that the proposed observer could estimate the states well in the presence of unknown inputs and missing measurements.

To further investigate the effects of the missing measurements on the estimation performance, observers with different λ are designed and their average

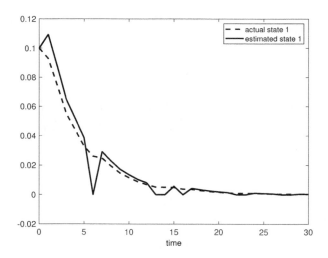

FIGURE 5.2: The state x_1 and its estimate

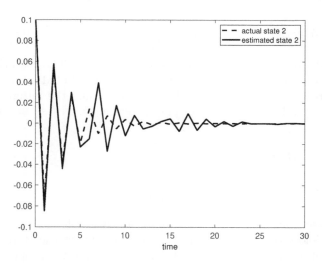

FIGURE 5.3: The state x_2 and its estimate

TABLE 5.1: Estimation Performance Comparison

λ	Average Estimation Error
0.15	1.006×10^{-1}
0.25	2.724×10^{-2}
0.35	1.213×10^{-2}
0.45	1.172×10^{-2}
0.55	9.259×10^{-3}
0.65	6.267×10^{-3}
0.75	4.152×10^{-3}
0.85	1.007×10^{-4}
0.95	1.638×10^{-5}

estimation errors obtained with 100 Monte Carlo experiments are presented in Table 5.1. It is noted that the one-step transition matrix F remains the same by properly adjusting K_1. We can conclude that better estimation results can be achieved when λ increases, which means that more useful information can lead to a better estimation performance.

5.5 Conclusions

The observer design problem has been considered for a class of systems with unknown inputs and missing measurements. A Bernoulli distributed sequence taking values on 0 or 1 has been introduced to govern possible data dropouts in the output transmission. An observer has been obtained with hope to decouple the estimation error from the unknown input in the mathematical expectation sense. The uniform ultimate boundedness of the estimation error has been investigated extensively as well. A simulation example has been provided to show the effectiveness of the proposed algorithm.

6

Filtering and Fault Detection for Nonlinear Systems with Polynomial Approximation

The past few decades have seen the nonlinear state estimation problem as a recurring research theme due to the pervasive existence of nonlinearities in almost all real-world industrial systems. If not adequately dealt with, the intrinsic nonlinearities may lead to undesirable dynamic behaviors such as oscillation or even instability. Indeed, the nonlinear analysis issue has been the main stream of research for systems control and estimation problems attracting researchers from a variety of disciplines. So far, much effort has been devoted to the estimation/filtering problems for nonlinear systems, see e.g. [49, 89] and the references therein. Among others, the renowned extended Kalman filter (EKF) algorithm has proved to be an effective method to solve the estimation problem for nonlinear systems in the least mean square sense. Recently, considerable attention has been paid to the performance improvement of the traditional EKF with respect to the insensitivity to the parameter uncertainties as well as the capability of handling nonlinearities [73, 136]. When the system states and observations are polynomial with additive Gaussian white noises, the mean-square filter has been designed where the statistical characteristics of Gaussian distribution have been made use of to recursively calculate the filtering error covariance.

Polynomial extended Kalman filter (PEKF) is an extension of EKF with aim to cater for inherent nonlinearies using polynomial approximations. Traditional EKF is only concerned with the *linear* term and simply ignores the linearization error, while PEKF considers the Carleman approximation of a nonlinear system of a given order μ [85]. The order could be determined according to the form of the nonlinearity and the estimation performance specification. In this sense, the PEKF is more applicable than EKF as far as the accuracy is concerned. When the order $\mu = 1$, PEKF reduces to conventional EKF. A PEKF is designed to cope with an augmented state which is made of Kronecker powers of the original state [52]. Due to its higher accuracy than that of EKF, the PEKF has stirred quite a lot research attention and many corresponding results have been reported in the literature [51]. Nevertheless, the PEKF approach still ignores the Carleman approximation errors which would give rise to certain conservatism especially when the nonlinearities are severe.

DOI: 10.1201/9781003309482-6

In theory, well-behaved nonlinear functions could only be approximated *accurately* by polynomials whose orders approach *infinity*. In engineering systems, however, the polynomials with extremely high orders are difficult to be realized in practice. A feasible way is to determine the finite order of the polynomials for satisfactory approximation of the nonlinear dynamics according to the degree of the nonlinearities and the nature of the research problem. As such, the unavoidable high-order approximation errors would be impacting on the estimation performances that should be taken into account. While EKF and PEKF work quite well for system with relatively low degree of nonlinearities, the classical EKF algorithm ignores the linearization errors and most available PEKF approaches discard the Carleman approximation errors. It is noted that the approximation errors differ greatly from each other due to various forms of nonlinearities. Instead of being simply dropped, the approximation errors do offer further room for improving the estimation accuracy if properly coped with. This seemingly natural idea, unfortunately, would inevitably bring us substantial challenges when calculating the covariances of the estimation errors in the least mean square sense since the approximation errors could not be exactly known. Moreover, coupled with both the low-order terms and external disturbances, the approximate errors would become very complicated to analyze. These identified difficulties motivate us to address the PEKF problem *by allowing for the Carleman approximation error* with aim to obtain higher approximation accuracy and better estimation performance.

The fault detection (FD) problem is another active research topic in control engineering due primarily to the increasingly higher and higher safety requirements. Yet, the FD problem provides an ideal platform to demonstrate the applicability of the polynomial filter technique to be developed. Since a properly designed filter could generate residual signal so as to efficiently detect abnormal changes in the system, filter/observer based FD has become a common technique. In filter/observer based FD methods, a fault would be detected and diagnosed effectively via comparing the actual system output with the filter output signal since the faults would normally bring in unexpected variations in the system measurement. To date, a great number of results have been published on this issue, see [62, 200] and the references therein. Residual evaluation, which consists of threshold and decision logic, is critical to the performance of FD. For systems varying with time or parameters, the *adaptive threshold* has been of particular research importance for its better trackability and faster self-adjustment compared with the constant threshold.

So far, many existing results have focused on adaptive threshold generation for linear systems. However, the corresponding results for nonlinear systems have been scattered in spite of their engineering significance. In [189, 193], the nonlinear dynamics has been assumed to be uniformly bounded and the bounds have been utilized to determine the adaptive threshold. In the context of polynomial fault detection for nonlinear systems, the adaptive threshold determination problem is essentially difficult mainly for two reasons: 1) the expressions of the disturbances and approximation errors are sophisticated and their influences on the threshold remain unclear; and 2) it is non-trivial to calculate the bounds of the disturbances and approximation errors in order to propose a reasonable fault detection threshold. Up to date, the polynomial fault detection scheme has not been fully investigated, not to mention the case where the Carleman approximation errors are taken into consideration. This constitutes another motivation of our present work.

Summarizing the discussions made above, in this chapter, both the polynomial filtering and fault detection problems are thoroughly investigated for a class of nonlinear systems. The Carleman approximation of a given order is introduced to approximate the nonlinear functions. In contrast to existing literature, the high-order approximation errors are explicitly taken into account in terms of low-order polynomials with uncertain but bounded coefficients. A time-varying filter is first designed at each time step to guarantee that the filtering error covariance is bounded in the fault-free case. Such a bound is subsequently minimized with respect to a properly designed filter gain. To show the applicability of the proposed filter scheme, a fault detection strategy is then proposed consisting of the calculation of adaptive threshold and decision logic. The filter gain and adaptive threshold are determined using the information from the approximation errors and additive disturbances.

The main contribution of the chapter can be highlighted as follows: 1) a polynomial estimation scheme for a class of nonlinear systems is presented by taking account of the Carleman approximation errors, which leads to higher estimation accuracy; 2) an upper bound of the estimation error covariance in the polynomial scheme is calculated and minimized by designing an appropriate filter gain; 3) adaptive fault detection threshold is developed for the desired polynomial filter to realize efficient detection and the fault detectability is analyzed; and 4) the proposed algorithms for both the filter gain design and the adaptive threshold computation are recursive and therefore suitable for online applications.

6.1 Problem Formulation

Consider the following class of stochastic discrete-time nonlinear systems:

$$\begin{cases} x_{k+1} = g(x_k, u_k) + v_k + f_k, \\ y_k = h(x_k) + w_k, \end{cases} \tag{6.1}$$

where $x_k \in \mathbb{R}^n$ is the system state; $y_k \in \mathbb{R}^b$ is the measurement output; $u_k \in \mathbb{R}^l$ is the control input; $f_k \in \mathbb{R}^n$ is the additive fault; $g : \mathbb{R}^n \times \mathbb{R}^l \mapsto \mathbb{R}^n$ and $h : \mathbb{R}^n \mapsto \mathbb{R}^b$ are $\mu + 1$ times continuously differentiable nonlinear maps. The state noise $v_k \in \mathbb{R}^n$ and the output noise $w_k \in \mathbb{R}^b$ are uncorrelated zero-mean sequences. The initial state x_0 is random and independent of the noises.

It is assumed that
$$\mathbb{E}\left\{x_0^{[i]}\right\} = \zeta_{0,i},\ \mathbb{E}\left\{v_k^{[i]}\right\} = \xi_{v,k,i},\ \mathbb{E}\left\{w_k^{[i]}\right\} = \xi_{w,k,i}, \quad (6.2)$$

for all $i = 1, 2, \ldots, 2\mu$, where $\zeta_{0,i}$, $\xi_{v,k,i}$ and $\xi_{w,k,i}$ are known vectors. Meanwhile, $x_0^{[i]}$, $v_k^{[i]}$, and $w_k^{[i]}$ are assumed to be distributed in bounded domains and the following inequalities hold definitely:

$$\left\|x_0^{[i]}\right\| \le s_{0,i},\ \left\|v_k^{[i]}\right\| \le s_{v,k,i},\ \left\|w_k^{[i]}\right\| \le s_{w,k,i}, \quad (6.3)$$

where $s_{0,i}$, $s_{v,k,i}$ and $s_{w,k,i}$ are known scalars for all $i = 1, 2, \ldots, \mu$. The fault f_k is assumed to be norm-bounded so that the measurements and the estimated states will not go infinity in a finite time in the faulty case.

6.1.1 Polynomial Approximation of Nonlinear Functions

Consider the sequences $x_{k+1}^{[m]}$ and $y_k^{[m]}$, the Kronecker powers of the state and measurement around the state estimate $\tilde{x}_k$ for $m = 0, 1, \ldots, \mu$. With Taylor polynomials around $\tilde{x}_k$, when the nonlinear functions are $\mu + 1$ times continuously differentiable, we have

$$\begin{cases} x_{k+1}^{[m]} = \sum_{i=0}^{\mu} G_{m,i}(\tilde{x}_k, u_k, v_k, f_k)(x_k - \tilde{x}_k)^{[i]} \\ \qquad + G_{m,\mu+1}(x_k, \tilde{x}_k, u_k, v_k, f_k)(x_k - \tilde{x}_k)^{[\mu+1]}, \\ y_k^{[m]} = \sum_{i=0}^{\mu} H_{m,i}(\tilde{x}_k, w_k)(x_k - \tilde{x}_k)^{[i]} + H_{m,\mu+1}(x_k, \tilde{x}_k, w_k)(x_k - \tilde{x}_k)^{[\mu+1]}, \end{cases}$$
(6.4)

where $\tilde{x}_{\theta_k} = \Theta_k x_k + (I - \Theta_k)\tilde{x}_k$, $\Theta_k = \mathrm{diag}\{\theta_{1k}, \ldots, \theta_{nk}\}$, $\theta_{ik} \in [0,1]$ for all $i = 1, 2, \ldots, n$, and

$$G_{m,i}(\tilde{x}_k, u_k, v_k, f_k) = \frac{1}{i!}\left(\nabla_x^{[i]} \otimes (g + v + f)^{[m]}\right)\Big|_{(x=\tilde{x}_k, u=u_k, v=v_k, f=f_k)}, \quad (6.5)$$

$$H_{m,i}(\tilde{x}_k, w_k) = \frac{1}{i!}\left(\nabla_x^{[i]} \otimes (h + w)^{[m]}\right)\Big|_{(x=\tilde{x}_k, w=w_k)}, \quad (6.6)$$

where the operation $\nabla_x^{[i]} \otimes$ applied to a function $\chi(x)$ is defined as:

$$\nabla_x^{[0]} \otimes \chi = \chi, \quad (6.7)$$
$$\nabla_x^{[i+1]} \otimes \chi = \nabla_x \otimes \nabla_x^{[i]} \otimes \chi, i > 0, \quad (6.8)$$

with $\nabla_x = [\partial/\partial x_1, \ldots, \partial/\partial x_n]$.

In order to obtain a least squares filter and determine the adaptive threshold for fault detection, now we consider system (6.1) with $f_k = 0$. Firstly we focus on the expression of the remainder terms. Denote

$$\nabla_x^{[\mu]} \otimes (g + v)^{[m]} = [\varepsilon_{1,\mu,m}(x, u, v), \ldots, \varepsilon_{n^m,\mu,m}(x, u, v)]^T, \quad (6.9)$$

Problem Formulation

where $\varepsilon_{i,\mu,m}(x,u,v) : \mathbb{R}^n \times \mathbb{R}^l \times \mathbb{R}^n \mapsto \mathbb{R}^{n^\mu}$ for all $i = 1, \ldots, n^m$. It can be easily verified that

$$G_{m,\mu+1}(x_k, \tilde{x}_k, u_k, v_k, 0)(x_k - \tilde{x}_k)^{[\mu+1]} = \mathfrak{G}(x_k, \tilde{x}_k, u_k, v_k)(x_k - \tilde{x}_k)^{[\mu]}, \tag{6.10}$$

where

$$\mathfrak{G}(x_k, \tilde{x}_k, u_k, v_k) = \frac{1}{(\mu+1)!} \begin{bmatrix} (x_k - \tilde{x}_k)^T(\nabla_x^T \otimes \varepsilon_{1,\mu,m}^T(\tilde{x}_{\theta_k}, u_k, v_k)) \\ \vdots \\ (x_k - \tilde{x}_k)^T(\nabla_x^T \otimes \varepsilon_{n^m,\mu,m}^T(\tilde{x}_{\theta_k}, u_k, v_k)) \end{bmatrix}. \tag{6.11}$$

We assume that x_k is bounded in an ellipsoid of center $\tilde{x}_k$ and shape matrix E_k, i.e. $x_k = \tilde{x}_k + E_k z_k$, for some $z_k \in \mathbb{R}^n$ and $\|z_k\| \leq 1$. Since the nonlinear functions $\nabla_x^T \otimes \varepsilon_{i,\mu,m}^T(\tilde{x}_{\theta_k}, u_k, v_k)$ are assumed to be continuous, their norm reaches an extremum on the bounded ellipsoid domain, thus there exist constants $\bar{m}_{1,k}, \ldots, \bar{m}_{n^m,k}$, such that for $i = 1, 2, \ldots, n^m$,

$$\|\nabla_x^T \otimes \varepsilon_{i,\mu,m}^T(\tilde{x}_{\theta_k}, u_k, v_k)\| \leq \bar{m}_{i,k}. \tag{6.12}$$

With $x_k = \tilde{x}_k + E_k z_k$ and $\|z_k\| \leq 1$, it follows directly that

$$\|(x_k - \tilde{x}_k)^T(\nabla_x^T \otimes \varepsilon_{i,\mu,m}^T(\tilde{x}_{\theta_k}, u_k, v_k))\| \leq \bar{m}_{i,k}\|E_k\|, \tag{6.13}$$

therefore there exist vectors $\varsigma_{i,k} \in \mathbb{R}^{n^\mu}$ such that $\|\varsigma_{i,k}\| \leq 1$ and

$$(x_k - \tilde{x}_k)^T(\nabla_x^T \otimes \varepsilon_{i,\mu,m}^T(\tilde{x}_{\theta_k}, u_k, v_k)) = \bar{m}_{i,k}\|E_k\|\varsigma_{i,k}^T. \tag{6.14}$$

Denoting $\Xi_k = [\varsigma_{1,k}, \ldots, \varsigma_{n^m,k}]^T$, we have

$$\mathfrak{G}(\tilde{x}_{\theta_k}, u_k, v_k) = \frac{1}{(\mu+1)!}\|E_k\|\text{diag}\{\bar{m}_{1,k}, \ldots, \bar{m}_{n^m,k}\}\Xi_k. \tag{6.15}$$

From the definition of Ξ_k and $\|\varsigma_{i,k}\| \leq 1$, it is obvious that $\|\Xi_k\| \leq \sqrt{n^m}$. Therefore, denote

$$L_{g,m,k} = \frac{\sqrt{n^m}}{(\mu+1)!}\|E_k\|\text{diag}\{\bar{m}_{1,k}, \ldots, \bar{m}_{n^m,k}\}, \tag{6.16}$$

and then $\mathfrak{G}(x_k, \tilde{x}_k, u_k, v_k)$ can be written as

$$\mathfrak{G}(x_k, \tilde{x}_k, u_k, v_k) = L_{g,m,k}\Delta_{g,m,k}, \tag{6.17}$$

where $L_{g,m,k}$ and $\Delta_{g,m,k}$ are problem-dependent matrices, and $\|\Delta_{g,m,k}\| \leq 1$. Similarly, we can have that

$$H_{m,\mu+1}(x_k, \tilde{x}_k, w_k)(x_k - \tilde{x}_k)^{[\mu+1]} = L_{h,m,k}\Delta_{h,m,k}(x_k - \tilde{x}_k)^{[\mu]}, \tag{6.18}$$

where $L_{h,m,k}$ and $\Delta_{h,m,k}$ are problem-dependent matrices, and $\|\Delta_{h,m,k}\| \leq 1$ as well.

Remark 6.1 *The way of processing the high-order approximation errors is along the similar line of coping with the linearization errors in [13, 178] where only linear terms with norm-bounded matrix coefficients have been utilized to formulate the linearization errors. Different from [13, 178], in this chapter, polynomial terms of orders higher than μ, namely, the approximation errors, are formulated as terms of lower orders with norm-bounded matrix coefficients. The unknown bounded matrices $\Delta_{g,m,k}$ and $\Delta_{h,m,k}$ are functions of both x_k and $\tilde{x}_k$, and they are not expressed as $\Delta_{g,m,k}(x_k,\tilde{x}_k)$ and $\Delta_{h,m,k}(x_k,\tilde{x}_k)$ only for simplicity. Notice that the approximation errors have been ignored in previous PEKF studies, which may lead to unsatisfactory estimation performance especially when the encountered nonlinearities are severe. Obviously, the approximation errors have a great influence on the estimation accuracy, and the novel error formulation (6.18) helps to achieve a better approximation and estimation performance in the sequel, which will be demonstrated in the simulation.*

6.1.2 The Polynomial Nonlinear Systems

Before proceeding further, the following lemma is introduced.

Lemma 6.1 *[85] For any integer $h \geq 0$ and for any $a, b \in \mathbb{R}^n$, we have*

$$(a+b)^{[h]} = \sum_{k=0}^{h} M_h^k (a^{[k]} \otimes b^{[h-k]}), \tag{6.19}$$

with a set of suitably defined matrices $M_h^k \in \mathbb{R}^{n^h \times n^h}$ $(k = 0, 1, \ldots, h)$.

With Lemma 6.1, (6.17), (6.18) and the fact that

$$(A \otimes B)(C \otimes D) = (AC) \otimes (BD), \tag{6.20}$$

(6.4) in the fault-free case can be written as follows:

$$\begin{cases} x_{k+1}^{[m]} = \sum\limits_{j=0}^{\mu} A_{m,j,k} x_k^{[j]} + L_{g,m,k} \Delta_{g,m,k} \sum\limits_{j=0}^{\mu} \Upsilon_{j,k} x_k^{[j]} + v_{m,k}, \\ y_k^{[m]} = \sum\limits_{j=0}^{\mu} C_{m,j,k} x_k^{[j]} + L_{h,m,k} \Delta_{h,m,k} \sum\limits_{j=0}^{\mu} \Upsilon_{j,k} x_k^{[j]} + w_{m,k}, \end{cases} \tag{6.21}$$

where

$$A_{m,j,k} = \sum_{i=j}^{\mu} \sum_{p=0}^{m} \frac{1}{i!} M_m^p \left(\left(\nabla_x^{[i]} \otimes g^{[p]} \right) \otimes \xi_{v,k,m-p} \right) M_i^j \left(I_{n^j} \otimes (-\tilde{x}_k)^{[i-j]} \right), \tag{6.22}$$

$$C_{m,j,k} = \sum_{i=j}^{\mu} \sum_{p=0}^{m} \frac{1}{i!} M_m^p \left(\left(\nabla_x^{[i]} \otimes h^{[p]} \right) \otimes \xi_{w,k,m-p} \right) M_i^j \left(I_{n^j} \otimes (-\tilde{x}_k)^{[i-j]} \right), \tag{6.23}$$

Problem Formulation

$$v_{m,k} = \sum_{i=0}^{\mu}\sum_{p=0}^{m} \frac{1}{i!} M_m^p \left(\left(\left(\nabla_x^{[i]} \otimes g^{[p]} \right) (x_k - \tilde{x}_k)^{[i]} \right) \otimes \left(v_k^{[m-p]} - \xi_{v,k,m-p} \right) \right), \tag{6.24}$$

$$w_{m,k} = \sum_{i=0}^{\mu}\sum_{p=0}^{m} \frac{1}{i!} M_m^p \left(\left(\left(\nabla_x^{[i]} \otimes h^{[p]} \right) (x_k - \tilde{x}_k)^{[i]} \right) \otimes \left(w_k^{[m-p]} - \xi_{w,k,m-p} \right) \right), \tag{6.25}$$

$$\Upsilon_{j,k} = M_\mu^j \left(I_{n^j} \otimes (-\tilde{x}_k)^{[\mu-j]} \right). \tag{6.26}$$

Denote the following augmented state and measurement:

$$x_k^{(\mu)} = \begin{bmatrix} 1 \\ x_k \\ x_k^{[2]} \\ \vdots \\ x_k^{[\mu]} \end{bmatrix}, y_k^{(\mu)} = \begin{bmatrix} 1 \\ y_k \\ y_k^{[2]} \\ \vdots \\ y_k^{[\mu]} \end{bmatrix}, \tag{6.27}$$

and (6.21) can be written as:

$$\begin{cases} x_{k+1}^{(\mu)} = A_k x_k^{(\mu)} + L_{g,k} \Delta_{g,k} \Upsilon_k x_k^{(\mu)} + v_k^{(\mu)}, \\ y_k^{(\mu)} = C_k x_k^{(\mu)} + L_{h,k} \Delta_{h,k} \Upsilon_k x_k^{(\mu)} + w_k^{(\mu)}, \end{cases} \tag{6.28}$$

where

$$A_k = \begin{bmatrix} A_{0,0,k} & \cdots & A_{0,\mu,k} \\ A_{1,0,k} & \cdots & A_{1,\mu,k} \\ \vdots & \ddots & \vdots \\ A_{\mu,0,k} & \cdots & A_{\mu,\mu,k} \end{bmatrix}, \quad C_k = \begin{bmatrix} C_{0,0,k} & \cdots & C_{0,\mu,k} \\ C_{1,0,k} & \cdots & C_{1,\mu,k} \\ \vdots & \ddots & \vdots \\ C_{\mu,0,k} & \cdots & C_{\mu,\mu,k} \end{bmatrix},$$

$$v_k^{(\mu)} = (v_{0,k}^T, v_{1,k}^T, \ldots, v_{\mu,k}^T)^T, \quad w_k^{(\mu)} = (w_{0,k}^T, w_{1,k}^T, \ldots, w_{\mu,k}^T)^T, \tag{6.29}$$

$$\Upsilon_k = [\Upsilon_{0,k}, \Upsilon_{1,k}, \ldots, \Upsilon_{\mu,k}], \quad \Delta_{h,k} = \text{diag}\{\Delta_{h,0,k}, \Delta_{h,1,k}, \ldots, \Delta_{h,\mu,k}\},$$

$$L_{g,k} = \text{diag}\{L_{g,0,k}, L_{g,1,k}, \ldots, L_{g,\mu,k}\}, \quad L_{h,k} = \text{diag}\{L_{h,0,k}, L_{h,1,k}, \ldots, L_{h,\mu,k}\},$$

$$\Delta_{g,k} = \text{diag}\{\Delta_{g,0,k}, \Delta_{g,1,k}, \ldots, \Delta_{g,\mu,k}\}, \tag{6.30}$$

and it is straightforward to see that $\|\Delta_{g,k}\| \leq 1$ and $\|\Delta_{h,k}\| \leq 1$. It is noted that when $g(x_k, u_k)$ (respectively, $h(x_k)$) is linear, the parameter $L_{g,k}$ (respectively, $L_{h,k}$) is zero. From the facts that $\mathbb{E}\left\{v_k^{[i]}\right\} = \xi_{v,k,i}$ and $\mathbb{E}\left\{w_k^{[i]}\right\} = \xi_{w,k,i}$, it follows that $v_k^{(\mu)}$ and $w_k^{(\mu)}$ are both zero-mean. However, it is quite difficult to calculate $\mathbb{E}\left\{v_k^{(\mu)}(v_k^{(\mu)})^T\right\}$ and $\mathbb{E}\left\{w_k^{(\mu)}(w_k^{(\mu)})^T\right\}$ due to that $v_k^{(\mu)}$ and $w_k^{(\mu)}$ involve the system state, the state estimate, and the nonlinear dynamics.

6.1.3 The Filter and the Fault Detection Problems

For system (6.28), a recursive filter to be designed is of the following form:

$$\tilde{x}_{k+1}^{(\mu)} = A_k \tilde{x}_k^{(\mu)} + K_k \left(y_k^{(\mu)} - C_k \tilde{x}_k^{(\mu)} \right), \qquad (6.31)$$

where $\tilde{x}_k^{(\mu)}$ is the estimate of $x_k^{(\mu)}$ at time step k with $\tilde{x}_0^{(\mu)} = \mathbb{E}\left\{ x_0^{(\mu)} \right\}$. K_k is the filter gain to be determined.

We are now in a position to state the addressed polynomial filter and fault detection problems as follows. 1) We are interested in designing the filter gain K_k in (6.31) for the system (6.28) such that the resulting estimation error covariance is bounded and such a bound is subsequently minimized. 2) Based on the proposed filter design scheme, the associated fault detection problem is to generate a residual signal whose threshold is adaptively computed by reflecting the approximation errors and external disturbances.

Remark 6.2 *Based on the degrees of the nonlinearities and the engineering requirement, the integer μ could be chosen in advance. A bigger μ would lead to higher approximation accuracy at the cost of heavier computation burden. In this sense, the determination of μ should be made according to the trade-off between the estimation accuracy and the computation expense. Therefore, the polynomial filter to be developed is applicable in many nonlinear cases since the order could be selected adaptively. When $\mu = 1$ and the approximation errors are not considered, the filter to be designed reduces to the conventional EKF.*

Remark 6.3 *Though the polynomial approximation errors could be represented in a mathematical way, they remain indeterminate to the designers because of the unknown matrix coefficients. As a result, the accurate estimation error covariance could not be obtained. A natural alternative is to find an upper bound of the estimation error covariance and then minimize it by a properly designed filter gain K_k. This way, the feasibility of the algorithm is enhanced and the robustness with respect to the approximation errors is maintained. After the determination of K_k, an adaptive fault detection strategy would be established accordingly.*

Remark 6.4 *It is observed from (6.21)–(6.28) that, in the proposed filter (6.31), A_k and C_k are quite complicated since they are nonlinear functions of the estimated states, addressed nonlinear dynamics as well as the statistics of disturbances. Rather than $A\left(\tilde{x}_k^{(\mu)}\right)$ and $C\left(\tilde{x}_k^{(\mu)}\right)$, they are denoted by A_k and C_k for simplicity only. By no means are they similar with the known system matrices in the linear case. In the estimation process, these two parameters have to be updated at each time instant by fairly complicated computation.*

6.2 Polynomial Filter Design

The following lemmas are essential in establishing the main results.

Lemma 6.2 *[176] Given matrices A, H, E, and F with appropriate dimensions such that $FF^T \leq I$. Let X be a symmetric positive definite matrix and γ be an arbitrary positive constant such that $\gamma^{-1}I > EXE^T$. Then, the following inequality holds*

$$(A + HFE) X (A + HFE)^T \leq A \left(X^{-1} - \gamma E^T E\right)^{-1} A^T + \gamma^{-1} H H^T. \quad (6.32)$$

Lemma 6.3 *[66] For any two vectors $x, y \in \mathbb{R}^n$, the following inequality holds*

$$xy^T + yx^T \leq \varepsilon xx^T + \varepsilon^{-1} yy^T, \quad (6.33)$$

where $\varepsilon > 0$ is a scalar.

Denote the estimation error as

$$e_k = x_k^{(\mu)} - \tilde{x}_k^{(\mu)}, \quad (6.34)$$

and the estimation error covariance conditional on the observations $y_j (j = 0, \ldots, k)$ as

$$P_{k+1} = \mathbb{E}\left\{e_{k+1} e_{k+1}^T | y_0, \ldots, y_k\right\}. \quad (6.35)$$

For presentation convenience, $\tilde{x}_k^{(\mu)}$, e_k, and P_k are partitioned as follows:

$$\tilde{x}_k^{(\mu)} = \begin{bmatrix} \tilde{x}_{0,k}^{(\mu)} \\ \tilde{x}_{1,k}^{(\mu)} \\ \vdots \\ \tilde{x}_{\mu,k}^{(\mu)} \end{bmatrix}, e_k = \begin{bmatrix} e_{0,k} \\ e_{1,k} \\ \vdots \\ e_{\mu,k} \end{bmatrix}, P_k = \begin{bmatrix} P_{0,0,k} & P_{0,1,k} & \cdots & P_{0,\mu,k} \\ P_{1,0,k} & P_{1,1,k} & \cdots & P_{1,\mu,k} \\ \vdots & \vdots & \ddots & \vdots \\ P_{\mu,0,k} & P_{\mu,1,k} & \cdots & P_{\mu,\mu,k} \end{bmatrix}, \quad (6.36)$$

where $e_{i,k}, \tilde{x}_{i,k}^{(\mu)} \in \mathbb{R}^{n^i}$ and $P_{i,j,k} \in \mathbb{R}^{n^i \times n^j}$ for any $i, j = 0, 1, \ldots, \mu$. It can be seen that $P_{i,j,0} = \text{st}_{n^i,n^j}^{-1}(\zeta_{0,i+j}) - \zeta_{0,i}\zeta_{0,j}^T$, where $\text{st}_{a,b}^{-1}$ is the inverse of the stack operator defined as: for a vector $\rho = [\rho_1, \ldots, \rho_{a \times b}]$, $\text{st}_{a,b}^{-1}(\rho) = [\varrho_{ij}]_{a \times b}$, where $\varrho_{ij} = \rho_{(j-1)a+i}$.

The first goal of this section is to design a filter in the form of (6.31) for system (6.28) such that an upper bound of the covariance of the filtering error P_k can be provided and minimized. For this purpose, let us now deal with the covariance of the filtering error in the following lemma.

Lemma 6.4 *Denoting*

$$\check{A}_k = (A_k - K_k C_k) + \sqrt{2}\left[L_{g,k}, -K_k L_{h,k}\right] \begin{bmatrix} \frac{1}{\sqrt{2}} \Delta_{g,k}^T, \frac{1}{\sqrt{2}} \Delta_{h,k}^T \end{bmatrix}^T \Upsilon_k, \quad (6.37)$$

the covariance of the filtering error in (6.35) obeys the following recursion:

$$\begin{aligned}P_{k+1} =& \mathbb{E}\left\{\check{A}_k P_k \check{A}_k^T\right\} + 2\left[L_{g,k}, -K_k L_{h,k}\right] \\ &\times \mathbb{E}\left\{\left[\frac{1}{\sqrt{2}}\Delta_{g,k}^T, \frac{1}{\sqrt{2}}\Delta_{h,k}^T\right]^T \Upsilon_k \tilde{x}_k^{(\mu)} \left(\tilde{x}_k^{(\mu)}\right)^T \Upsilon_k^T \left[\frac{1}{\sqrt{2}}\Delta_{g,k}^T, \frac{1}{\sqrt{2}}\Delta_{h,k}^T\right]\right\} \\ &\times \left[L_{g,k}, -K_k L_{h,k}\right]^T + \sqrt{2}\mathbb{E}\left\{\check{A}_k e_k \left(\tilde{x}_k^{(\mu)}\right)^T \Upsilon_k^T \left[\frac{1}{\sqrt{2}}\Delta_{g,k}^T, \frac{1}{\sqrt{2}}\Delta_{h,k}^T\right]\right\} \\ &\times \left[L_{g,k}, -K_k L_{h,k}\right]^T + \sqrt{2}\left[L_{g,k}, -K_k L_{h,k}\right] \\ &\times \mathbb{E}\left\{\left[\frac{1}{\sqrt{2}}\Delta_{g,k}^T, \frac{1}{\sqrt{2}}\Delta_{h,k}^T\right]^T \Upsilon_k \tilde{x}_k^{(\mu)} e_k^T \check{A}_k^T\right\} + \Psi_k^v + K_k \Psi_k^w K_k^T,\end{aligned}$$

(6.38)

where $\Psi_k^v = \mathbb{E}\left\{v_k^{(\mu)}(v_k^{(\mu)})^T\right\}$ and $\Psi_k^w = \mathbb{E}\left\{w_k^{(\mu)}(w_k^{(\mu)})^T\right\}$.

Proof Substituting (6.28) and (6.31) into (6.34) yields

$$\begin{aligned}e_{k+1} =& \left((A_k - K_k C_k) + \sqrt{2}\left[L_{g,k}, -K_k L_{h,k}\right]\left[\frac{1}{\sqrt{2}}\Delta_{g,k}^T, \frac{1}{\sqrt{2}}\Delta_{h,k}^T\right]^T \Upsilon_k\right) e_k \\ &+ \sqrt{2}\left[L_{g,k}, -K_k L_{h,k}\right]\left[\frac{1}{\sqrt{2}}\Delta_{g,k}^T, \frac{1}{\sqrt{2}}\Delta_{h,k}^T\right]^T \Upsilon_k \tilde{x}_k^{(\mu)} + v_k^{(\mu)} - K_k w_k^{(\mu)}.\end{aligned}$$

(6.39)

Noticing that $v_k^{(\mu)}$ and $w_k^{(\mu)}$ are zero-mean disturbances and independent of e_k, (6.38) can be obtained directly with the definition of $\check{A}_k$. The proof now is complete.

As discussed before, the accurate estimation error covariance in Lemma 6.4 could not be obtained due to the unknown matrix coefficients. To solve the estimation problem, an upper bound of the estimation error covariance will be derived in the next theorem, and the filter gain will then be designed to minimize the bound at each time step.

Theorem 6.1 *Let $\gamma_{1,k}, \gamma_{2,k}, \gamma_{3,k}, \varepsilon$ be positive scalars. With initial condition $\bar{P}_0 = P_0$, assume that the following discrete Riccati-like equation*

$$\begin{aligned}\bar{P}_{k+1} =& (1+\varepsilon)A_k \left(\bar{P}_k^{-1} - \gamma_{1,k}\Upsilon_k^T \Upsilon_k\right)^{-1} A_k^T \\ &+ 2\left(\gamma_{1,k}^{-1}(1+\varepsilon) + \gamma_{2,k}^{-1}(1+\varepsilon^{-1})\right) L_{g,k} L_{g,k}^T + \bar{\Psi}_k^v - Z_k^T Y_k^{-1} Z_k \end{aligned}$$ (6.40)

has positive definite solutions such that the following constraints

$$\gamma_{1,k}^{-1} I - \Upsilon_k \bar{P}_k \Upsilon_k^T > 0, \tag{6.41}$$

$$\gamma_{2,k}^{-1} I - \Upsilon_k \tilde{x}_k^{(\mu)}(\tilde{x}_k^{(\mu)})^T \Upsilon_k^T > 0, \tag{6.42}$$

$$\gamma_{3,k}^{-1} I - \Upsilon_k \bar{X}_k \Upsilon_k^T > 0, \tag{6.43}$$

Polynomial Filter Design

are satisfied, where

$$Y_k = (1+\varepsilon)C_k \left(\bar{P}_k^{-1} - \gamma_{1,k}\Upsilon_k^T\Upsilon_k\right)^{-1} C_k^T$$
$$+ 2\left(\gamma_{1,k}^{-1}(1+\varepsilon) + \gamma_{2,k}^{-1}(1+\varepsilon^{-1})\right) L_{h,k}L_{h,k}^T + \bar{\Psi}_k^w, \quad (6.44)$$

$$Z_k = (1+\varepsilon)C_k \left(\bar{P}_k^{-1} - \gamma_{1,k}\Upsilon_k^T\Upsilon_k\right)^{-1} A_k^T, \quad (6.45)$$

$$\bar{\Psi}_{a,b,k}^v = \sum_{i=0}^{\mu}\sum_{j=0}^{\mu}\sum_{p=0}^{a}\sum_{q=0}^{b} \frac{1}{i!j!} M_a^p \Bigg(\left(\left(\nabla_x^{[i]} \otimes g^{[p]}\right) \bar{\Omega}_{i,j,k} \left(\nabla_x^{[j]} \otimes g^{[q]}\right)^T\right) \quad (6.46)$$

$$\otimes \left(\mathrm{st}_{n^{a-p},n^{b-q}}^{-1}(\xi_{v,k,a-p+b-q}) - \xi_{v,k,a-p}\xi_{v,k,b-q}^T\right)\Bigg)(M_b^q)^T,$$

$$\bar{\Psi}_{a,b,k}^w = \sum_{i=0}^{\mu}\sum_{j=0}^{\mu}\sum_{p=0}^{a}\sum_{q=0}^{b} \frac{1}{i!j!} M_a^p \Bigg(\left(\left(\nabla_x^{[i]} \otimes h^{[p]}\right) \bar{\Omega}_{i,j,k} \left(\nabla_x^{[j]} \otimes h^{[q]}\right)^T\right)$$
$$(6.47)$$

$$\otimes \left(\mathrm{st}_{n^{a-p},n^{b-q}}^{-1}(\xi_{w,k,a-p+b-q}) - \xi_{w,k,a-p}\xi_{w,k,b-q}^T\right)\Bigg)(M_b^q)^T,$$

$$\bar{\Omega}_{i,j,k} = \sum_{p=0}^{i}\sum_{q=0}^{j} M_i^p \Bigg(\left(I_{n^p} \otimes (-\tilde{x}_k)^{[i-p]}\right) \bar{X}_{p,q,k} \left(I_{n^q} \otimes (-\tilde{x}_k)^{[j-q]}\right)^T\Bigg)(M_j^q)^T,$$
$$(6.48)$$

$$\bar{X}_{k+1} = A_k \left(\bar{X}_k^{-1} - \gamma_{3,k}\Upsilon_k^T\Upsilon_k\right)^{-1} A_k^T + \gamma_{3,k}^{-1} L_{g,k}L_{g,k}^T + \bar{\Psi}_k^v, \quad (6.49)$$

$$\bar{X}_0 = \mathbb{E}\left\{x_0^{(\mu)}\left(x_0^{(\mu)}\right)^T\right\}.$$

Similar with P_k, matrices $\bar{X}_k$, $\bar{P}_k$, $\bar{\Psi}_k^v$ and $\bar{\Psi}_k^w$ are partitioned as follows:

$$\bar{X}_k = \begin{bmatrix} \bar{X}_{0,0,k} & \bar{X}_{0,1,k} & \cdots & \bar{X}_{0,\mu,k} \\ \bar{X}_{1,0,k} & \bar{X}_{1,1,k} & \cdots & \bar{X}_{1,\mu,k} \\ \vdots & \vdots & \ddots & \vdots \\ \bar{X}_{\mu,0,k} & \bar{X}_{\mu,1,k} & \cdots & \bar{X}_{\mu,\mu,k} \end{bmatrix}, \bar{P}_k = \begin{bmatrix} \bar{P}_{0,0,k} & \bar{P}_{0,1,k} & \cdots & \bar{P}_{0,\mu,k} \\ \bar{P}_{1,0,k} & \bar{P}_{1,1,k} & \cdots & \bar{P}_{1,\mu,k} \\ \vdots & \vdots & \ddots & \vdots \\ \bar{P}_{\mu,0,k} & \bar{P}_{\mu,1,k} & \cdots & \bar{P}_{\mu,\mu,k} \end{bmatrix},$$

$$\bar{\Psi}_k^v = \begin{bmatrix} \bar{\Psi}_{0,0,k}^v & \bar{\Psi}_{0,1,k}^v & \cdots & \bar{\Psi}_{0,\mu,k}^v \\ \bar{\Psi}_{1,0,k}^v & \bar{\Psi}_{1,1,k}^v & \cdots & \bar{\Psi}_{1,\mu,k}^v \\ \vdots & \vdots & \ddots & \vdots \\ \bar{\Psi}_{\mu,0,k}^v & \bar{\Psi}_{\mu,1,k}^v & \cdots & \bar{\Psi}_{\mu,\mu,k}^v \end{bmatrix}, \bar{\Psi}_k^w = \begin{bmatrix} \bar{\Psi}_{0,0,k}^w & \bar{\Psi}_{0,1,k}^w & \cdots & \bar{\Psi}_{0,\mu,k}^w \\ \bar{\Psi}_{1,0,k}^w & \bar{\Psi}_{1,1,k}^w & \cdots & \bar{\Psi}_{1,\mu,k}^w \\ \vdots & \vdots & \ddots & \vdots \\ \bar{\Psi}_{\mu,0,k}^w & \bar{\Psi}_{\mu,1,k}^w & \cdots & \bar{\Psi}_{\mu,\mu,k}^w \end{bmatrix},$$
$$(6.50)$$

where $\bar{X}_{i,j,k}, \bar{P}_{i,j,k}, \bar{\Psi}_{i,j,k}^v \in \mathbb{R}^{n^i \times n^j}$, $\bar{\Psi}_{i,j,k}^w \in \mathbb{R}^{m^i \times m^j}$ for any $i, j = 0, 1, \ldots, \mu$. Then, with the filter gain given by

$$K_k = Z_k^T Y_k^{-1}, \quad (6.51)$$

the matrix $\bar{P}_k$ is an upper bound of P_k in the fault-free case. Moreover, the filter gain given by (6.51) minimizes the upper bound at each time step.

Proof For notational simplicity, we denote

$$\Psi^v_{i,j,k} = \mathbb{E}\left\{v_{i,k}v_{j,k}^T\right\}, \quad \Psi^w_{i,j,k} = \mathbb{E}\left\{w_{i,k}w_{j,k}^T\right\}, \quad X_k = \mathbb{E}\left\{x_k^{(\mu)}\left(x_k^{(\mu)}\right)^T\right\},$$

$$\Omega_{i,j,k} = \mathbb{E}\left\{(x_k - \tilde{x}_k)^{[i]}\left((x_k - \tilde{x}_k)^{[j]}\right)^T\right\}, \quad (6.52)$$

and Ψ^v_k and Ψ^w_k have been defined in Lemma 6.4.

Before proving that $\bar{P}_k$ is an upper bound of P_k, we are going to show that $X_k \leq \bar{X}_k$, $\Psi^v_k \leq \bar{\Psi}^v_k$, and $\Psi^w_k \leq \bar{\Psi}^w_k$. These results could be obtained by induction. With $\bar{X}_0 = \mathbb{E}\left\{x_0^{(\mu)}\left(x_0^{(\mu)}\right)^T\right\}$, we assume that $i = 1, 2, \ldots, k$, $X_i \leq \bar{X}_i$. Now it remains to show that $\Psi^v_k \leq \bar{\Psi}^v_k$, $\Psi^w_k \leq \bar{\Psi}^w_k$, and $X_{k+1} \leq \bar{X}_{k+1}$.

Noting the fact that v_k is independent of x_k, it follows from Lemma 6.1, (6.20), and (6.24) that

$$\Psi^v_{a,b,k} = \sum_{i=0}^{\mu}\sum_{j=0}^{\mu}\sum_{p=0}^{a}\sum_{q=0}^{b}\frac{1}{i!j!}M_a^p\Bigg(\left(\left(\nabla_x^{[i]}\otimes g^{[p]}\right)\Omega_{i,j,k}\left(\nabla_x^{[j]}\otimes g^{[q]}\right)^T\right)$$

$$\otimes \mathbb{E}\left\{\left(v^{[a-p]} - \xi_{v,k,a-p}\right)\left(v^{[b-q]} - \xi_{v,k,b-q}\right)^T\right\}\Bigg)(M_b^q)^T, \quad (6.53)$$

and

$$\Omega_{i,j,k} = \sum_{p=0}^{i}\sum_{q=0}^{j}M_i^p\bigg(\left(I_{n^p}\otimes(-\tilde{x}_k)^{[i-p]}\right)\mathbb{E}\left\{x_k^{[p]}\left(x_k^{[q]}\right)^T\right\}$$

$$\times \left(I_{n^q}\otimes(-\tilde{x}_k)^{[j-q]}\right)^T\bigg)(M_j^q)^T. \quad (6.54)$$

In fact, $v_{m,k}$ can be written as:

$$v_{m,k} = \sum_{i=0}^{\mu}\sum_{p=0}^{m}\frac{1}{i!}M_m^p\left(\left(\nabla_x^{[i]}\otimes g^{[p]}\right)\otimes\left(v_k^{[m-p]} - \xi_{v,k,m-p}\right)\right)(x_k - \tilde{x}_k)^{[i]}.$$

$$(6.55)$$

Recalling Lemma 6.1 and the definition of $x_k^{(\mu)}$, one has

$$v_{m,k} = \mathfrak{X}_m(\tilde{x}_k, u_k, v_k)x_k^{(\mu)}, \quad (6.56)$$

where $\mathfrak{X}_m(\cdot,\cdot,\cdot)$ is a proper nonlinear mapping. Then, $v_k^{(\mu)}$ can be expressed as

$$v_k^{(\mu)} = \mathfrak{X}(\tilde{x}_k, u_k, v_k)x_k^{(\mu)}, \quad (6.57)$$

Polynomial Filter Design

where $\mathfrak{X}(\tilde{x}_k, u_k, v_k) = \begin{bmatrix} \mathfrak{X}_0^T(\tilde{x}_k, u_k, v_k), \mathfrak{X}_1^T(\tilde{x}_k, u_k, v_k), \ldots, \mathfrak{X}_\mu^T(\tilde{x}_k, u_k, v_k) \end{bmatrix}^T$. Now, it follows from the assumption $X_k \leq \bar{X}_k$ and the independence between v_k and x_k that

$$\Psi_k^v = \mathbb{E}\left\{\mathfrak{X}(\tilde{x}_k, u_k, v_k)\mathbb{E}\left\{x_k^{(\mu)}\left(x_k^{(\mu)}\right)^T\right\}\mathfrak{X}^T(\tilde{x}_k, u_k, v_k)\right\}$$
$$\leq \mathbb{E}\left\{\mathfrak{X}(\tilde{x}_k, u_k, v_k)\bar{X}_k\mathfrak{X}^T(\tilde{x}_k, u_k, v_k)\right\}. \qquad (6.58)$$

The above inequality shows that, by replacing $\mathbb{E}\left\{x_k^{[p]}\left(x_k^{[q]}\right)^T\right\}$ with $\bar{X}_{p,q,k}$ in (6.54), we would get an upper bound of Ψ_k^v. Therefore, from (6.46) and (6.48), it follows directly that $\Psi_k^v \leq \bar{\Psi}_k^v$. Similarly, we can get $\Psi_k^w \leq \bar{\Psi}_k^w$.

It remains to show that $X_{k+1} \leq \bar{X}_{k+1}$. With (6.28), we have

$$X_{k+1} = \mathbb{E}\left\{(A_k + L_{g,k}\Delta_{g,k}\Upsilon_k)X_k(A_k + L_{g,k}\Delta_{g,k}\Upsilon_k)^T\right\} + \Psi_k^v. \qquad (6.59)$$

Since it has been proved that $\Psi_k^v \leq \bar{\Psi}_k^v$, we have

$$X_{k+1} \leq \mathbb{E}\left\{(A_k + L_{g,k}\Delta_{g,k}\Upsilon_k)X_k(A_k + L_{g,k}\Delta_{g,k}\Upsilon_k)^T\right\} + \bar{\Psi}_k^v. \qquad (6.60)$$

With the assumption that $X_k \leq \bar{X}_k$ and the condition (6.43), it follows from Lemma 6.2 that

$$X_{k+1} \leq A_k\left(\bar{X}_k^{-1} - \gamma_{3,k}\Upsilon_k^T\Upsilon_k\right)^{-1}A_k^T + \gamma_{3,k}^{-1}L_{g,k}L_{g,k}^T + \bar{\Psi}_k^v = \bar{X}_{k+1}. \qquad (6.61)$$

So far, we have proved that $X_k \leq \bar{X}_k$, $\Psi_k^v \leq \bar{\Psi}_k^v$, and $\Psi_k^w \leq \bar{\Psi}_k^w$, and therefore we are going to deal with P_k next. The corresponding results can also be obtained by induction. It is already known that $\bar{P}_0 = P_0$. Then, assuming that, for $i = 1, 2, \ldots, k$, $P_i \leq \bar{P}_i$ and it remains to prove that $P_{k+1} \leq \bar{P}_{k+1}$.

With Lemma 6.3, we can have the following inequality:

$$\sqrt{2}\check{A}_k e_k\left(\tilde{x}_k^{(\mu)}\right)^T\Upsilon_k^T\left[\frac{1}{\sqrt{2}}\Delta_{g,k}^T, \frac{1}{\sqrt{2}}\Delta_{h,k}^T\right][L_{g,k}, -K_kL_{h,k}]^T \qquad (6.62)$$
$$+ \sqrt{2}[L_{g,k}, -K_kL_{h,k}]\left[\frac{1}{\sqrt{2}}\Delta_{g,k}^T, \frac{1}{\sqrt{2}}\Delta_{h,k}^T\right]^T\Upsilon_k\tilde{x}_k^{(\mu)}e_k^T\check{A}_k^T$$
$$\leq \varepsilon\check{A}_k e_k e_k^T\check{A}_k^T + 2\varepsilon^{-1}[L_{g,k}, -K_kL_{h,k}]\left[\frac{1}{\sqrt{2}}\Delta_{g,k}^T, \frac{1}{\sqrt{2}}\Delta_{h,k}^T\right]^T$$
$$\times \Upsilon_k\tilde{x}_k^{(\mu)}\left(\tilde{x}_k^{(\mu)}\right)^T\Upsilon_k^T\left[\frac{1}{\sqrt{2}}\Delta_{g,k}^T, \frac{1}{\sqrt{2}}\Delta_{h,k}^T\right][L_{g,k}, -K_kL_{h,k}]^T.$$

Substituting (6.62) into (6.38) yields

$$P_{k+1} \leq (1+\varepsilon)\mathbb{E}\left\{\check{A}_k P_k \check{A}_k^T\right\} + 2(1+\varepsilon^{-1})\left[L_{g,k}, -K_k L_{h,k}\right]$$
$$\times \mathbb{E}\left\{\left[\frac{1}{\sqrt{2}}\Delta_{g,k}^T, \frac{1}{\sqrt{2}}\Delta_{h,k}^T\right]^T \Upsilon_k \tilde{x}_k^{(\mu)} \left(\tilde{x}_k^{(\mu)}\right)^T \Upsilon_k^T \right.$$
$$\left. \times \left[\frac{1}{\sqrt{2}}\Delta_{g,k}^T, \frac{1}{\sqrt{2}}\Delta_{h,k}^T\right]\right\} \left[L_{g,k}, -K_k L_{h,k}\right]^T + \Psi_k^v + K_k \Psi_k^w K_k^T. \quad (6.63)$$

From the upper bounds of the covariances of $v_k^{(\mu)}$ and $w_k^{(\mu)}$ and the assumption that $P_k \leq \bar{P}_k$, we have

$$P_{k+1} \leq (1+\varepsilon)\mathbb{E}\left\{\check{A}_k \bar{P}_k \check{A}_k^T\right\} + 2(1+\varepsilon^{-1})\left[L_{g,k}, -K_k L_{h,k}\right]$$
$$\times \mathbb{E}\left\{\left[\frac{1}{\sqrt{2}}\Delta_{g,k}^T, \frac{1}{\sqrt{2}}\Delta_{h,k}^T\right]^T \Upsilon_k \tilde{x}_k^{(\mu)} \left(\tilde{x}_k^{(\mu)}\right)^T \Upsilon_k^T \right.$$
$$\left. \times \left[\frac{1}{\sqrt{2}}\Delta_{g,k}^T, \frac{1}{\sqrt{2}}\Delta_{h,k}^T\right]\right\} \left[L_{g,k}, -K_k L_{h,k}\right]^T + \bar{\Psi}_k^v + K_k \bar{\Psi}_k^w K_k^T. \quad (6.64)$$

Noticing that $\left\|\left[\frac{1}{\sqrt{2}}\Delta_{f,k}^T, \frac{1}{\sqrt{2}}\Delta_{h,k}^T\right]\right\| \leq 1$, it follows from (6.41), (6.42), and Lemma 6.2 that

$$P_{k+1} \leq (1+\varepsilon)(A_k - K_k C_k)\left(\bar{P}_k^{-1} - \gamma_{1,k}\Upsilon_k^T \Upsilon_k\right)^{-1}(A_k - K_k C_k)^T$$
$$+ 2\gamma_{1,k}^{-1}(1+\varepsilon)\left(L_{g,k}L_{g,k}^T + K_k L_{h,k}L_{h,k}^T K_k^T\right)$$
$$+ 2\gamma_{2,k}^{-1}(1+\varepsilon^{-1})\left(L_{g,k}L_{g,k}^T + K_k L_{h,k}L_{h,k}^T K_k^T\right) + \bar{\Psi}_k^v + K_k \bar{\Psi}_k^w K_k^T. \quad (6.65)$$

Furthermore, it can be seen from (6.44) and (6.45) that

$$P_{k+1} \leq K_k Y_k K_k^T - K_k Z_k - Z_k^T K_k^T + (1+\varepsilon)A_k\left(\bar{P}_k^{-1} - \gamma_{1,k}\Upsilon_k^T \Upsilon_k\right)^{-1} A_k^T$$
$$+ 2\left(\gamma_{1,k}^{-1}(1+\varepsilon) + \gamma_{2,k}^{-1}(1+\varepsilon^{-1})\right)L_{g,k}L_{g,k}^T + \bar{\Psi}_k^v. \quad (6.66)$$

Noticing $Y_k = Y_k^T > 0$ and completing the square with respect to K_k, we have

$$P_{k+1} \leq \left(K_k - Z_k^T Y_k^{-1}\right) Y_k \left(K_k - Z_k^T Y_k^{-1}\right)^T - Z_k^T Y_k^{-1} Z_k$$
$$+ (1+\varepsilon)A_k\left(\bar{P}_k^{-1} - \gamma_{1,k}\Upsilon_k^T \Upsilon_k\right)^{-1} A_k^T$$
$$+ 2\left(\gamma_{1,k}^{-1}(1+\varepsilon) + \gamma_{2,k}^{-1}(1+\varepsilon^{-1})\right)L_{g,k}L_{g,k}^T + \bar{\Psi}_k^v. \quad (6.67)$$

It is now obvious that, when $K_k = Z_k^T Y_k^{-1}$, the upper bound of P_{k+1} is minimized and

$$P_{k+1} \leq -Z_k^T Y_k^{-1} Z_k + (1+\varepsilon)A_k\left(\bar{P}_k^{-1} - \gamma_{1,k}\Upsilon_k^T \Upsilon_k\right)^{-1} A_k^T$$
$$+ 2\left(\gamma_{1,k}^{-1}(1+\varepsilon) + \gamma_{2,k}^{-1}(1+\varepsilon^{-1})\right)L_{g,k}L_{g,k}^T + \bar{\Psi}_k^v = \bar{P}_{k+1}, \quad (6.68)$$

which concludes the proof.

Fault Detection

Remark 6.5 *The filter gain K_k is calculated at each time instant to minimize an upper bound of the filtering error covariance. The system (6.28) under consideration includes the polynomial approximation errors, thereby better reflecting the reality. Specific efforts have been made to handle the approximation errors. Though the unknown matrices $\Delta_{g,k}$ and $\Delta_{h,k}$ can not be introduced to design the filter, the matrices $L_{g,k}$ and $L_{h,k}$ reflect the effects of the approximation errors on the filter design in a quantitative way. The value of the state estimate $\tilde{x}_k^{(\mu)}$ can be directly used to design K_k since $\tilde{x}_k^{(\mu)}$ is already obtained at time step k. Due to the consideration of the polynomial approximation errors, the accurate covariances Ψ_k^v and Ψ_k^w cannot be obtained as done in [51,52]. Instead, the upper bounds of the covariances have been calculated and employed to design the suboptimal filter. It is worth mentioning that, $\bar{\Phi}_k^w$, an upper bound of $\mathbb{E}\left\{w_k^{(\mu)}(w_k^{(\mu)})^T\right\}$, could guarantee the invertibility of Y_k in (6.51). The parameter ε can be determined to balance the intrinsic characteristic of the proposed polynomial filter and the impact brought from the state estimates in the upper bound of the filtering error covariance. The proposed algorithm can be applied to nonlinear time-varying systems that are $\mu + 1$ times continuously differentiable. Furthermore, the gain K_k can be designed recursively by a Riccati-like equation, which is applicable for online computations.*

Theorem 6.1 provides a recursive way to obtain the filter gain K_k. The algorithm is summarized as follows to show the calculation of K_k at each time step with a given order μ.

Algorithm 6.1

Step 1. Select the initial values $\bar{P}_0$ and $\tilde{x}_0^{(\mu)}$ based on the distribution of x_0.

Step 2. Determine E_k and $\bar{m}_{i,k}(i = 1, \ldots, n^\mu)$ according to the system dynamics, the disturbances, and the initial estimation error, and compute the Carleman approximation matrices A_k, C_k, $L_{g,k}$, $L_{h,k}$, and Υ_k based on (6.16)–(6.18) and (6.21)–(6.29).

Step 3. Determine $\gamma_{1,k}$, $\gamma_{2,k}$, and $\gamma_{3,k}$ with (6.41)–(6.43).

Step 4. Compute $\bar{\Omega}_k$, $\bar{\Phi}_k^w$, and $\bar{\Phi}_k^v$ according to (6.46)–(6.50).

Step 5. Choose a positive scalar ε, and calculate Y_k and Z_k based on (6.44) and (6.45).

Step 6. Obtain the desired filter gain K_k with (6.51).

Step 7. Update the estimate $\tilde{x}_{k+1}^{(\mu)}$ by (6.31).

Step 8. Calculate $\bar{P}_{k+1}$ and $\bar{X}_{k+1}$ using (6.40) and (6.49).

Step 9. Set $k = k + 1$ and go to Step 2.

6.3 Fault Detection

To efficiently detect the fault in the system (6.1), we need to define a residual signal based on the proposed polynomial filter and then evaluate it properly. Here, we utilize the following signal as the residual:

$$r_k = y_k^{(\mu)} - C_k \tilde{x}_k^{(\mu)}. \tag{6.69}$$

In the next theorem, a time-varying fault detection threshold will be calculated for the residual defined in (6.69). The state estimate, approximation error, and bounds of the noises are all taken into consideration. Such a threshold would improve the sensitivity of the fault detection.

Theorem 6.2 *For system (6.1), let us consider the filter (6.31) designed in Theorem 6.1. Then, the residual in (6.69) satisfies the following inequality in the fault free case:*

$$\|r_k\| \leq (\|C_k\| + \|L_{h,k}\| \|\Upsilon_k\|) \delta_{e,k} + \|L_{h,k}\| \left\|\Upsilon_k \tilde{x}_k^{(\mu)}\right\| + \delta_{w,k} := \delta_{r,k}, \tag{6.70}$$

where

$$\delta_{e,k+1} = \left(\|A_k - K_k C_k\| + \sqrt{2} \|[L_{g,k}, -K_k L_{h,k}]\| \|\Upsilon_k\|\right) \delta_{e,k} \tag{6.71}$$
$$+ \sqrt{2} \|[L_{g,k}, -K_k L_{h,k}]\| \left\|\Upsilon_k \tilde{x}_k^{(\mu)}\right\| + \delta_{v,k} + \|K_k\| \delta_{w,k},$$

$$\delta_{v,m,k} = \sum_{i=0}^{\mu} \sum_{p=0}^{m} \frac{1}{i!} \|M_m^p\| \left\|\nabla_x^{[i]} \otimes g^{[p]}\right\| \bar{\rho}_{i,k} \left(s_{v,k,m-p} + \|\xi_{v,k,m-p}\|\right), \tag{6.72}$$

$$\delta_{w,m,k} = \sum_{i=0}^{\mu} \sum_{p=0}^{m} \frac{1}{i!} \|M_m^p\| \left\|\nabla_x^{[i]} \otimes h^{[p]}\right\| \bar{\rho}_{i,k} \left(s_{w,k,m-p} + \|\xi_{w,k,m-p}\|\right), \tag{6.73}$$

$$\bar{\rho}_{i,k} = \sum_{q=0}^{i} \left\|M_i^q \left(I_{n^q} \otimes (-\tilde{x}_k)^{[i-q]}\right)\right\| \left(\left\|\tilde{x}_{q,k}^{(\mu)}\right\| + \delta_{e,k}\right), \tag{6.74}$$

$$\delta_{v,k} = \sqrt{\sum_{m=0}^{\mu} (\delta_{v,m,k})^2}, \tag{6.75}$$

$$\delta_{w,k} = \sqrt{\sum_{m=0}^{\mu} (\delta_{w,m,k})^2}, \tag{6.76}$$

and the initial values of $\delta_{e,k}$ are given by

$$\delta_{e,i,0} = s_{0,i} + \|\zeta_{0,i}\|, \delta_{e,0} = \sqrt{\sum_{i=0}^{\mu} \delta_{e,i,0}^2}. \tag{6.77}$$

Then, $\delta_{r,k}$ is the threshold that we can utilize to detect the fault.

Proof *Substituting (6.28) and (6.31) into (6.69), we have*

$$r_k = C_k e_k + L_{h,k} \Delta_{h,k} \Upsilon_k (e_k + \tilde{x}_k^{(\mu)}) + w_k^{(\mu)}. \tag{6.78}$$

Fault Detection 105

With $\|\Delta_{h,k}\| \leq 1$ and triangle inequality, we can get

$$\|r_k\| \leq (\|C_k\| + \|L_{h,k}\|\|\Upsilon_k\|)\|e_k\| + \|L_{h,k}\|\|\Upsilon_k \tilde{x}_k^{(\mu)}\| + \|w_k^{(\mu)}\|. \quad (6.79)$$

If we can prove that $\|e_k\| \leq \delta_{e,k}$ and $\|w_k^{(\mu)}\| \leq \delta_{w,k}$, then (6.70) follows directly from (6.79).

It follows from (6.25) that,

$$\|w_{m,k}^{(\mu)}\| \leq \sum_{i=0}^{\mu}\sum_{p=0}^{m} \frac{1}{i!} \|M_m^p\| \left\|\nabla_x^{[i]} \otimes h^{[p]}\right\| \left\|(x_k - \tilde{x}_k)^{[i]}\right\| \left\|w_k^{[m-p]} - \xi_{w,k,m-p}\right\|, \quad (6.80)$$

and the upper bound of $\left\|(x_k - \tilde{x}_k)^{[i]}\right\|$ can be determined as follows:

$$\left\|(x_k - \tilde{x}_k)^{[i]}\right\| = \left\|\sum_{q=0}^{i} M_i^q \left(I_{n^q} \otimes (-\tilde{x}_k)^{[i-q]}\right) x_k^{[q]}\right\|$$

$$\leq \sum_{q=0}^{i} \left\|M_i^q \left(I_{n^q} \otimes (-\tilde{x}_k)^{[i-q]}\right)\right\| \left\|\tilde{x}_{q,k}^{(\mu)} + e_{q,k}\right\|. \quad (6.81)$$

Noticing the fact that $\|e_{q,k}\| \leq \|e_k\|$, we have

$$\left\|(x_k - \tilde{x}_k)^{[i]}\right\| \leq \sum_{q=0}^{i} \left\|M_i^q \left(I_{n^q} \otimes (-\tilde{x}_k)^{[i-q]}\right)\right\| \left(\|e_k\| + \|\tilde{x}_k^{(\mu)}\|\right). \quad (6.82)$$

With (6.82), we can see that if $\|e_k\| \leq \delta_{e,k}$, then $\|w_{m,k}^{(\mu)}\| \leq \delta_{w,m,k}$ and $\|w_k^{(\mu)}\| \leq \delta_{w,k}$. Similarly, we have that if $\|e_k\| \leq \delta_{e,k}$, then $\|v_k^{(\mu)}\| \leq \delta_{v,k}$. Thus, to obtain (6.70), all we need to verify is that $\|e_k\| \leq \delta_{e,k}$. This can be proved by induction. It is obvious that $\|e_0\| \leq \delta_{e,0}$. Assuming that $\|e_i\| \leq \delta_{e,i}$ ($i = 1, 2, \ldots, k$), we are going to demonstrate that $\|e_{k+1}\| \leq \delta_{e,k+1}$.

With (6.39) and $\left\|\left[\frac{1}{\sqrt{2}}\Delta_{f,k}^T, \frac{1}{\sqrt{2}}\Delta_{h,k}^T\right]\right\| \leq 1$, it follows that

$$\|e_{k+1}\| \leq \left(\|A_k - K_k C_k\| + \sqrt{2}\|[L_{g,k}, -K_k L_{h,k}]\|\|\Upsilon_k\|\right)\|e_k\| \quad (6.83)$$
$$+ \sqrt{2}\|[L_{g,k}, -K_k L_{h,k}]\|\left\|\Upsilon_k \tilde{x}_k^{(\mu)}\right\| + \|v_k^{(\mu)}\| + \|K_k\|\|w_k^{(\mu)}\|.$$

Based on our assumption that $\|e_k\| \leq \delta_{e,k}$, we have $\|w_k^{(\mu)}\| \leq \delta_{w,k}$ and $\|v_k^{(\mu)}\| \leq \delta_{v,k}$. Then, it is straightforward to see that $\|e_{k+1}\| \leq \delta_{e,k+1}$. Now we can draw the conclusion that $\|e_k\| \leq \delta_{e,k}$. Substituting $\|e_k\| \leq \delta_{e,k}$ and $\|w_k^{(\mu)}\| \leq \delta_{w,k}$ into (6.79) yields (6.70). The proof now is complete.

The algorithm is summarized as follows to show the determination of the threshold $\delta_{r,k}$ at each time step with a given order μ.

Algorithm 6.2

Step 1. Determine the initial values $\delta_{r,0}$ with (6.70) and (6.77).
Step 2. $\|A_k\|$, $\|C_k\|$, $\|L_{g,k}\|$, $\|L_{h,k}\|$, $\|\Upsilon_k\|$, and $\|K_k\|$ based on (6.21)-(6.28) and (6.51).
Step 3. Compute $\bar{\rho}_{i,k}$ with (6.74).
Step 4. Choose $\delta_{v,k}$ and $\delta_{w,k}$ according to (6.72),(6.73), (6.75), and (6.76).
Step 5. Obtain $\delta_{r,k}$ based on (6.70).
Step 6. Update $\delta_{e,k+1}$ based on (6.71).
Step 7. Set $k = k+1$ and go to Step 2.

Based on Theorem 6.2, a natural fault detection strategy follows directly as follows:

$$\begin{cases} \|r_k\| \leq \delta_{r,k} & \implies \text{no fault} \\ \|r_k\| > \delta_{r,k} & \implies \text{fault detected} \end{cases} \quad (6.84)$$

With the proposed filter and fault detection threshold, the fault detectability is analyzed in the sequel.

Theorem 6.3 *For system (6.1), consider the filter (6.31) designed in Theorem 6.1 and the fault detection strategy given in (6.84). The fault will be detected at time step $k+1$ if the following inequality holds:*

$$\alpha_{v,f,k} > \frac{1}{|\|C_{k+1}\| - \|L_{f,h,k+1}\|\|\Upsilon_{k+1}\||} \quad (6.85)$$

$$\times \left(\delta_{r,k+1} + \delta_{w,k+1} + \|L_{f,h,k+1}\|\|\Upsilon_{k+1}\tilde{x}_{k+1}^{(\mu)}\|\right) + \|K_k\|\|w_k^{(\mu)}\|$$

$$+ \left(\|A_k - K_k C_k\| + \sqrt{2}\|[L_{f,g,k}, -K_k L_{f,h,k}]\|\|\Upsilon_k\|\right)\delta_{e,k}$$

$$+ \sqrt{2}\|[L_{f,g,k}, -K_k L_{f,h,k}]\|\left\|\Upsilon_k \tilde{x}_k^{(\mu)}\right\|,$$

where

$$\alpha_{v,f,m,k} = (\|f_k\| - s_{v,k,1})^m - \|\xi_{v,k,m}\| - \sum_{i=0}^{\mu}\sum_{p=1}^{m}\frac{1}{i!}\left\|M_m^p\left(\nabla_x^{[i]} \otimes g^{[p]}\right)\right\|\bar{\rho}_{i,k}$$

$$\times \left((\|f_k\| + s_{v,k,1})^{m-p} + \|\xi_{v,k,m-p}\|\right), \quad (6.86)$$

$$\alpha_{v,f,k} = \sqrt{\sum_{m=0}^{\mu}(\alpha_{v,f,m,k})^2}, \quad (6.87)$$

$L_{f,g,k}$ and $L_{f,h,k}$ are the problem-dependent matrices in the faulty case, which are corresponding to $L_{g,k}$ and $L_{h,k}$ in the fault-free case, respectively.

Fault Detection

Proof In the presence of f_k, $v_{m,k}$ can be written as

$$v_{m,k} = \sum_{i=0}^{\mu} \sum_{p=0}^{m} \frac{1}{i!} M_m^p \left(\left(\left(\nabla_x^{[i]} \otimes g^{[p]} \right) \left(x_k - \tilde{x}_k \right)^{[i]} \right) \right.$$
$$\left. \otimes \left((v_k + f_k)^{[m-p]} - \xi_{v,k,m-p} \right) \right). \tag{6.88}$$

Then it follows directly that

$$\|v_{m,k}\| \geq (\|f_k\| - s_{v,k,1})^m - \|\xi_{v,k,m}\| - \sum_{i=0}^{\mu} \sum_{p=1}^{m} \frac{1}{i!} \left\| M_m^p \left(\nabla_x^{[i]} \otimes g^{[p]} \right) \right\|$$
$$\times \bar{\rho}_{i,k} \left((\|f_k\| + s_{v,k,1})^{m-p} + \|\xi_{v,k,m-p}\| \right)$$
$$= \alpha_{v,f,m,k}. \tag{6.89}$$

Subsequently, we have

$$\|v_k^{(\mu)}\| \geq \alpha_{v,f,k}. \tag{6.90}$$

With (6.39), we have

$$\|e_{k+1}\| \geq \|v_k^{(\mu)}\| - \left(\|A_k - K_k C_k\| + \sqrt{2} \|[L_{f,g,k}, -K_k L_{f,h,k}]\| \|\Upsilon_k\| \right) \|e_k\|$$
$$- \sqrt{2} \|[L_{f,g,k}, -K_k L_{f,h,k}]\| \left\| \Upsilon_k \tilde{x}_k^{(\mu)} \right\| - \|K_k\| \|w_k^{(\mu)}\|. \tag{6.91}$$

Substituting (6.85) into (6.91) yields

$$\|e_{k+1}\| > \frac{1}{\|\|C_{k+1}\| - \|L_{f,h,k+1}\| \|\Upsilon_{k+1}\|\|}$$
$$\times \left(\delta_{r,k+1} + \delta_{w,k+1} + \|L_{f,h,k+1}\| \|\Upsilon_{k+1} \tilde{x}_{k+1}^{(\mu)}\| \right). \tag{6.92}$$

Considering (6.78), we have

$$\|r_{k+1}\| \geq \|\|C_{k+1}\| - \|L_{f,h,k+1}\| \|\Upsilon_{k+1}\|\| \|e_{k+1}\| - \|L_{f,h,k+1}\| \|\Upsilon_{k+1} \tilde{x}_{k+1}^{(\mu)}\|$$
$$- \|w_{k+1}^{(\mu)}\|. \tag{6.93}$$

From (6.92) and (6.93), it follows that

$$\|r_{k+1}\| \geq \delta_{r,k+1}. \tag{6.94}$$

Based on the proposed fault detection strategy (6.84), the fault can be detected at time step $k+1$ and that concludes the proof.

Remark 6.6 *Theorem 6.2 puts forward a recursive method to determine the fault detection threshold for system (6.28). With the adaptive threshold, abnormal changes in systems can be detected accordingly. The fault may be some state-dependent abrupt changes (e.g. actuator or sensor failures in a control system). The approximation errors, additive disturbances, and initial estimation error have all been considered in the threshold. The spectral norms of matrices and the upper bounds of Euclidean norms of vector disturbances have been introduced to calculate the threshold. In this way, the false alarms can be avoided in the proposed fault detection scheme. Moreover, the threshold is related to the measurement and estimated states, and hence needs to be updated online at each time step, which makes the method adaptive. With the proposed filter and fault detection strategy, the fault detectability is analyzed in Theorem 6.3 as well. The fault detectability condition established in Theorem 6.3 takes the Carleman approximation and the approximation error into account and therefore looks complicated. Nevertheless, such a condition can be easily verified and also provides inequality constraints on the fault size. The explicit consideration of the approximation error in the filter and fault detector designs constitutes the main difference between our work and those in [51, 52], and the derivation of the threshold and fault detectability analysis for nonlinear systems with polynomial approximation are new. If the stability analysis of the proposed filter approach becomes a concern, some additional assumptions can be made on the system parameters so as to ensure the boundedness of the estimation errors [83, 136]. In the next section, a numerical simulation would be carried out to show the effectiveness of the presented fault detection strategy.*

6.4 Illustrative Example

Consider the following nonlinear discrete system:

$$\begin{cases} x_{1,k+1} = 0.85x_{1,k} + 0.5x_{2,k}\sin(x_{1,k}) + v_{1,k} + f_k, \\ x_{2,k+1} = 1.15u_k - 0.5x_{1,k}\sin(x_{2,k}) + v_{2,k}, \\ y_k = x_{2,k} + w_k, \\ u_k = y_k. \end{cases}$$

The state noises and measurement noise are independent and obey the following distributions (for $i = 1, 2$):

$$\begin{cases} P\left(v_{i,k} = -1 \times 10^{-3}\right) = 0.6, \\ P\left(v_{i,k} = 0\right) = 0.2, \\ P\left(v_{i,k} = 3 \times 10^{-3}\right) = 0.2, \end{cases}$$

Illustrative Example   109

and
$$\begin{cases} P\left(w_k = -7 \times 10^{-3}\right) = 0.3, \\ P\left(w_k = 3 \times 10^{-3}\right) = 0.7. \end{cases}$$

When the order $\mu = 2$, how to construct the polynomial filter is explained step by step as follows. For the addressed system, we have

$$\begin{cases} g(x_k, u_k) = \begin{bmatrix} 0.85 x_{1,k} + 0.5 x_{2,k} \sin(x_{1,k}) \\ 1.15 u_k - 0.5 x_{1,k} \sin(x_{2,k}) \end{bmatrix} \\ h(x_k) = x_{2,k}, \end{cases}$$

With the linear measurement, we can easily obtain that $L_{h,k} = 0$ and

$$C_k = \begin{bmatrix} 1 & 0\ 0\ 0\ 0\ 0\ 0 \\ 0 & 0\ 1\ 0\ 0\ 0\ 0 \\ \mathbb{E}\{w_k^{[2]}\} & 0\ 0\ 0\ 0\ 0\ 1 \end{bmatrix}.$$

With (6.22) and (6.29), A_k can be determined as follows:

$A_{0,0,k} = 1,$
$A_{0,1,k} = [0\ 0],$
$A_{0,2,k} = [0\ 0\ 0\ 0],$
$A_{1,0,k} = \begin{bmatrix} -0.25 \tilde{x}_{1,k}^2 \tilde{x}_{2,k} \sin(\tilde{x}_{1,k}) \\ 1.15 u_k + 0.25 \tilde{x}_{1,k} \tilde{x}_{2,k}^2 \sin(\tilde{x}_{2,k}) \end{bmatrix},$
$A_{1,1,k} = \begin{bmatrix} 0.85 + 0.5 \tilde{x}_{1,k} \tilde{x}_{2,k} \sin(\tilde{x}_{1,k}) & 0.5 \sin(\tilde{x}_{1,k}) - 0.5 \tilde{x}_{1,k} \cos(\tilde{x}_{1,k}) \\ -0.5 \sin(\tilde{x}_{2,k}) + 0.5 \tilde{x}_{2,k} \cos(\tilde{x}_{2,k}) & -0.5 \tilde{x}_{1,k} \tilde{x}_{2,k} \sin(\tilde{x}_{2,k}) \end{bmatrix},$
$A_{1,2,k} = \begin{bmatrix} \pi_{01k} & 0.25 \cos(\tilde{x}_{1,k}) & 0.25 \cos(\tilde{x}_{1,k}) & 0 \\ 0 & -0.25 \cos(\tilde{x}_{2,k}) & -0.25 \cos(\tilde{x}_{2,k}) & \pi_{02k} \end{bmatrix},$
$A_{2,0,k} = \xi_{v,k,2} + \begin{bmatrix} \pi_{11k} & \pi_{12k} & \pi_{12k} & \pi_{13k} \end{bmatrix}^T,$
$A_{2,1,k} = \begin{bmatrix} \pi_{21k} & \pi_{22k} & \pi_{22k} & \pi_{23k} \\ \pi_{24k} & \pi_{25k} & \pi_{25k} & \pi_{26k} \end{bmatrix}^T,$
$A_{2,2,k} = \begin{bmatrix} \pi_{31k} & \pi_{32k} & \pi_{32k} & \pi_{33k} \\ \pi_{34k} & \pi_{35k} & \pi_{35k} & \pi_{36k} \\ \pi_{34k} & \pi_{35k} & \pi_{35k} & \pi_{36k} \\ \pi_{37k} & \pi_{38k} & \pi_{38k} & \pi_{39k} \end{bmatrix},$

where

$\pi_{01k} = -0.25 \tilde{x}_{2,k} \sin(\tilde{x}_{1,k})$
$\pi_{02k} = 0.25 \tilde{x}_{1,k} \sin(\tilde{x}_{2,k})$
$\pi_{11k} = 0.25 \tilde{x}_{1,k} \tilde{x}_{2,k}^2 \sin(\tilde{x}_{1,k}) \cos(\tilde{x}_{1,k}) - 0.2125 \tilde{x}_{1,k}^3 \tilde{x}_{2,k} \sin(\tilde{x}_{1,k})$
$\qquad + 0.125 \tilde{x}_{1,k}^2 \tilde{x}_{2,k}^2 (\cos^2(\tilde{x}_{1,k}) - \sin^2(\tilde{x}_{1,k})) + 0.425 \tilde{x}_{1,k}^2 \tilde{x}_{2,k} \cos(\tilde{x}_{1,k}),$

$$\begin{aligned}
\pi_{12k} =& -0.425\tilde{x}_{1,k}^2\tilde{x}_{2,k}\cos(\tilde{x}_{2,k}) - 0.14375u_k\tilde{x}_{1,k}^2\tilde{x}_{2,k}\sin(\tilde{x}_{1,k}) \\
& - 0.25\tilde{x}_{1,k}^2\tilde{x}_{2,k}\cos(\tilde{x}_{1,k})\sin(\tilde{x}_{2,k}) + 0.125\tilde{x}_{1,k}^3\tilde{x}_{2,k}\sin(\tilde{x}_{1,k})\sin(\tilde{x}_{2,k}) \\
& - 0.25\tilde{x}_{1,k}^2\tilde{x}_{2,k}^2\cos(\tilde{x}_{1,k})\cos(\tilde{x}_{2,k}) + 0.2125\tilde{x}_{1,k}^2\tilde{x}_{2,k}^2\sin(\tilde{x}_{2,k}) \\
& - 0.25\tilde{x}_{1,k}\tilde{x}_{2,k}^2\sin(\tilde{x}_{1,k})\cos(\tilde{x}_{2,k}) + 0.125\tilde{x}_{1,k}\tilde{x}_{2,k}^3\sin(\tilde{x}_{1,k})\sin(\tilde{x}_{2,k}),\\
\pi_{13k} =& 1.3225u_k^2 + 0.5\tilde{x}_{1,k}^2\tilde{x}_{2,k}\sin(\tilde{x}_{2,k})\cos(\tilde{x}_{2,k}) + 0.55u_k\tilde{x}_{1,k}\tilde{x}_{2,k}^2\sin(\tilde{x}_{2,k}) \\
& + 0.25\tilde{x}_{1,k}^2\tilde{x}_{2,k}^2(\cos^2(\tilde{x}_{2,k}) - \sin^2(\tilde{x}_{2,k})),\\
\pi_{21k} =& -0.5\tilde{x}_{2,k}^2\sin(\tilde{x}_{1,k})\cos(\tilde{x}_{1,k}) - 1.7\tilde{x}_{1,k}\tilde{x}_{2,k}\cos(\tilde{x}_{1,k}) + 0.85\tilde{x}_{1,k}^2\tilde{x}_{2,k} \\
& \times \sin(\tilde{x}_{1,k}) - 0.5\tilde{x}_{1,k}\tilde{x}_{2,k}^2(\cos^2(\tilde{x}_{1,k}) - \sin^2(\tilde{x}_{1,k})),\\
\pi_{22k} =& 0.9775u_k + 0.575u_k\tilde{x}_{1,k}\tilde{x}_{2,k}\sin(\tilde{x}_{1,k}) + 0.5\tilde{x}_{1,k}\tilde{x}_{2,k}\cos(\tilde{x}_{1,k})\sin(\tilde{x}_{2,k}) \\
& - 0.25\tilde{x}_{1,k}^2\tilde{x}_{2,k}\sin(\tilde{x}_{1,k})\sin(\tilde{x}_{2,k}) + 0.85\tilde{x}_{1,k}\tilde{x}_{2,k}\cos(\tilde{x}_{2,k}) \\
& + 0.25\tilde{x}_{1,k}^2\sin(\tilde{x}_{1,k})\cos(\tilde{x}_{2,k}) + 0.25\tilde{x}_{1,k}\tilde{x}_{2,k}^2\cos(\tilde{x}_{1,k})\cos(\tilde{x}_{2,k}),\\
\pi_{23k} =& -1.15u_k\sin(\tilde{x}_{2,k}) + 1.15u_k\tilde{x}_{2,k}\cos(\tilde{x}_{2,k}) - \tilde{x}_{1,k}\tilde{x}_{2,k}\sin(\tilde{x}_{2,k})\cos(\tilde{x}_{2,k}),\\
\pi_{24k} =& -0.85\tilde{x}_{1,k}^2\cos(\tilde{x}_{1,k}) - \tilde{x}_{1,k}\tilde{x}_{2,k}\sin(\tilde{x}_{1,k})\cos(\tilde{x}_{1,k}),\\
\pi_{25k} =& 0.575u_k\sin(\tilde{x}_{1,k}) + 0.425\tilde{x}_{1,k}^2\cos(\tilde{x}_{2,k}) - 0.575u_k\tilde{x}_{1,k}\cos(\tilde{x}_{1,k}) \\
& + 0.25\tilde{x}_{1,k}^2\cos(\tilde{x}_{1,k})\sin(\tilde{x}_{2,k}) + 0.25\tilde{x}_{1,k}^2\tilde{x}_{2,k}\cos(\tilde{x}_{1,k})\cos(\tilde{x}_{2,k}) \\
& - 0.425\tilde{x}_{1,k}^2\tilde{x}_{2,k}\sin(\tilde{x}_{2,k}) + 0.5\tilde{x}_{1,k}\tilde{x}_{2,k}\sin(\tilde{x}_{1,k})\cos(\tilde{x}_{2,k}) \\
& - 0.25\tilde{x}_{1,k}\tilde{x}_{2,k}^2\sin(\tilde{x}_{1,k})\sin(\tilde{x}_{2,k}),\\
\pi_{26k} =& -0.5\tilde{x}_{1,k}^2\sin(\tilde{x}_{2,k})\cos(\tilde{x}_{2,k}) - 1.15u_k\tilde{x}_{1,k}\tilde{x}_{2,k}\sin(\tilde{x}_{2,k}) \\
& - 0.5\tilde{x}_{1,k}^2\tilde{x}_{2,k}(\cos^2(\tilde{x}_{2,k}) - \sin^2(\tilde{x}_{2,k})),\\
\pi_{31k} =& 0.7225 + 0.85\tilde{x}_{2,k}\cos(\tilde{x}_{1,k}) - 0.425\tilde{x}_{1,k}\tilde{x}_{2,k}\sin(\tilde{x}_{1,k}) \\
& + 0.25\tilde{x}_{2,k}^2(\cos^2(\tilde{x}_{1,k}) - \sin^2(\tilde{x}_{1,k})),\\
\pi_{32k} =& 0.425\sin(\tilde{x}_{1,k}) + 0.425\tilde{x}_{1,k}\cos(\tilde{x}_{1,k}) + 0.5\tilde{x}_{2,k}\sin(\tilde{x}_{1,k})\cos(\tilde{x}_{1,k}),\\
\pi_{33k} =& 0.25\sin^2(\tilde{x}_{1,k}),\\
\pi_{34k} =& -0.2875u_k\tilde{x}_{2,k}\sin(\tilde{x}_{1,k}) - 0.425\sin(\tilde{x}_{2,k}) - 0.25\tilde{x}_{2,k}\cos(\tilde{x}_{1,k})\sin(\tilde{x}_{2,k}) \\
& + 0.125\tilde{x}_{1,k}\tilde{x}_{2,k}\sin(\tilde{x}_{1,k})\sin(\tilde{x}_{2,k}),\\
\pi_{35k} =& 0.2875u_k\cos(\tilde{x}_{1,k}) - 0.425\tilde{x}_{1,k}\cos(\tilde{x}_{2,k}) - 0.125\sin(\tilde{x}_{1,k})\sin(\tilde{x}_{2,k}) \\
& - 0.125\tilde{x}_{2,k}\sin(\tilde{x}_{1,k})\cos(\tilde{x}_{2,k}) - 0.125\tilde{x}_{1,k}\cos(\tilde{x}_{1,k})\sin(\tilde{x}_{2,k}) \\
& - 0.125\tilde{x}_{1,k}\tilde{x}_{2,k}\cos(\tilde{x}_{1,k})\cos(\tilde{x}_{2,k}),\\
\pi_{36k} =& 0.2125\tilde{x}_{1,k}^2\sin(\tilde{x}_{2,k}) - 0.25\tilde{x}_{1,k}\sin(\tilde{x}_{1,k})\cos(\tilde{x}_{2,k}) \\
& + 0.125\tilde{x}_{1,k}\tilde{x}_{2,k}\sin(\tilde{x}_{1,k})\sin(\tilde{x}_{2,k}),\\
\pi_{37k} =& 0.25\sin^2(\tilde{x}_{2,k}),\\
\pi_{38k} =& -0.575u_k\cos(\tilde{x}_{2,k}) + 0.5\tilde{x}_{1,k}\sin(\tilde{x}_{2,k})\cos(\tilde{x}_{2,k}),\\
\pi_{39k} =& 0.575u_k\tilde{x}_{1,k}\sin(\tilde{x}_{2,k}) + 0.25\tilde{x}_{1,k}^2(\cos^2(\tilde{x}_{2,k}) - \sin^2(\tilde{x}_{2,k})).
\end{aligned}$$

Illustrative Example

TABLE 6.1: Average Trace of the Estimation Error Covariance

	$\mu = 1$	$\mu = 2$
Traditional PEKF	6.1763×10^{-4}	4.9531×10^{-4}
Proposed Filter	5.5649×10^{-4}	3.9534×10^{-4}

To this end, A_k can be obtained.

Based on (6.26), we have

$$\Upsilon_k = \begin{bmatrix} \tilde{x}_{1,k}^2 & -2\tilde{x}_{1,k} & 0 & 1 & 0 & 0 & 0 \\ \tilde{x}_{1,k}\tilde{x}_{2,k} & -\tilde{x}_{2,k} & -\tilde{x}_{1,k} & 0 & 1 & 0 & 0 \\ \tilde{x}_{1,k}\tilde{x}_{2,k} & -\tilde{x}_{2,k} & -\tilde{x}_{1,k} & 0 & 0 & 1 & 0 \\ \tilde{x}_{2,k}^2 & 0 & -2\tilde{x}_{2,k} & 0 & 0 & 0 & 1 \end{bmatrix}.$$

The matrix E_k is set to be $0.075I$ based on the system dynamics, the disturbances, and the initial estimation error, and it can be seen that the assumption on the estimation error (i.e. $x_k = \tilde{x}_k + E_k z_k$ for some $z_k \in \mathbb{R}^n$ and $\|z_k\| \leq 1$) holds with such a choice of E_k. Based on E_k and the values of $\bar{m}_{i,k}$ for $i = 1, 2, \ldots, n^\mu$, the matrix $L_{g,k}$ can be chosen as $0.1I$. After getting all the aforementioned parameters, Theorem 6.1 can be adopted to recursively calculate the desired filter gain.

In the fault-free case, the average estimation error covariances with the proposed method and those with the conventional PEKF algorithm [51, 52] are both listed in Table 6.1. Monte-Carlo simulations with 50 runs are carried out in the cases where $\mu = 1$ and $\mu = 2$. From the table, two conclusions can be drawn: 1) approximating the system with a polynomial of a higher order implies a more accurate approximation of the original nonlinear system and gives better estimation results, no matter whether the approximation errors are considered or not; and 2) the proposed filter outperforms classical PEKF, thanks to the specific efforts we have made to deal with the high-order approximation errors. To demonstrate the fault detection strategy, we add the following fault to the system:

$$f_k = \begin{cases} 0.003, & \text{if } k > 20, \\ 0, & \text{otherwise.} \end{cases}$$

Figs. 6.1 and 6.2 show the actual states and the estimates when $\mu = 2$. It can be concluded that, the proposed filter performs well when there is no fault in the system, and the obvious difference between the actual state and estimated state in the faulty case could help us to detect the fault. The Euclidean norm of residual and the adaptive threshold are illustrated in Fig. 6.3. It can be seen that using the adaptive threshold, the additive fault could be detected immediately after it occurs. This result presents the effectiveness of our adaptive fault detection strategy.

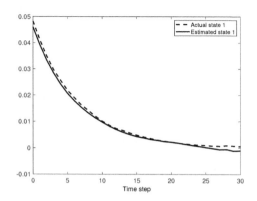

FIGURE 6.1: The state x_1 and its estimate

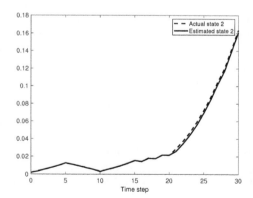

FIGURE 6.2: The state x_2 and its estimate

Conclusion

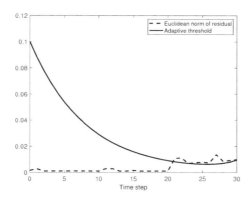

FIGURE 6.3: The Euclidean norm of residual and the threshold

6.5 Conclusion

The filtering and fault detection problems for a class of nonlinear systems have been addressed with a polynomial approach in the chapter. The nonlinear functions have been approximated with polynomials of a chosen degree, and the approximation errors, which result from high-order remainder terms of series expansions, have been written as low-order terms with norm-bounded coefficients. An upper bound of the filtering error covariance has been derived in the presence of the approximation errors. Then, the filter has been designed to minimize the bound at each time step in the fault-free case. The desired filter gain could be obtained with a set of Riccati-like recursive matrix equations, and thus the algorithm is applicable for online computation. A fault detection strategy with adaptive threshold has been proposed to efficiently detect the possible fault, taking account of the noises and the approximation errors. Sufficient conditions have been established to guarantee the fault detectability for the proposed fault detection scheme. A numerical simulation has been given to illustrate that the proposed method can achieve accurate state estimation and effective fault detection. It should be noted that two of the future research topics would be to investigate the fault diagnosis and fault tolerant control problems after the fault detection.

7

Event-Triggered Filtering and Fault Estimation for Nonlinear Systems with Stochastic Sensor Saturations

In the past few decades, the event-triggered transmission (ETT) mechanism has aroused a great deal of interest due to the rapid development of computer science and digital microprocessor. Compared with the conventional clock-driven strategy referring to periodic signal transmissions, in an ETT scheme, the outputs/inputs are released only when some conditions are violated. By reducing signal exchanges, the ETT could avoid some harmful transmission phenomena (e.g. data dropout, time delay, and congestion), improve the energy efficiency, and extend the lifetime of the services.

Recently, the event-triggered filtering (ETF) problem has started to gain some initial research attention especially for systems with wireless links and energy constraints. For example, the event-triggered H_∞ filtering problem with transmission delays has been investigated in [67] and a modified Kalman filter for linear systems with event-triggered transmissions has been designed in [153] where the differences between the measurements have been assumed to be uniformly distributed. In [172], the event-triggered minimum-variance filter has been thoroughly studied where the probability density functions (PDFs) of the states and the innovations conditional on measurements have been approximated with a sum of Gaussian distributions. However, when the system model is relatively complicated, the conditional PDFs will be intrinsically non-Gaussian and the Gaussian approximations may be quite inaccurate. Therefore, there appears to be a practical need to develop an alternative approach for addressing the ETF problem without strong assumptions on the distribution of measurements.

Due to physical and technological limitations, sensors/actuators cannot provide signals with unbounded amplitudes and such saturation phenomena pose extra challenges to the systems design. The control/filtering problems with actuator/sensor saturations have drawn much research attention where most available literature has treated the saturations as sector-bounded non-linearities. Nevertheless, sensors in practical systems might frequently encounter some transient phenomena especially when systems are deployed in changeable environments such as power grids [57, 82]. Under the circumstances, the saturation itself may undergo random switches/changes in its occurrence/intensity because of various reasons such as random sensor

116 Event-Triggered Filtering and Fault Estimation for Nonlinear Systems

failures and abrupt environmental changes [164]. As such, it would be interesting to examine the impact of both the ETT and stochastic saturations on the filter performance in the minimum variance sense. Note that the filtering problem with stochastic saturations has not received adequate research attention yet, not to mention the case when the nonlinearity and ETT are also taken into account. Note that, 1) it is novel to cope with the ETT issue without the approximated conditional PDFs of states and innovations; and 2) it would be non-trivial to include the saturation level and the statistical characteristics of the sensor saturations in the filter design.

In this chapter, we aim to solve the filtering problem for a class of nonlinear systems subject to event-triggered measurement transmissions and stochastic sensor saturations. Some Bernoulli-distributed sequences are introduced to govern the stochastic sensor saturations. By introducing the thresholds in the event generator and the saturation levels, an upper bound of the filtering error covariance is obtained. The filter gain is properly designed with hope to minimize the upper bound at each time step. The filtering performance is analyzed with respect to the error boundedness. Sufficient conditions are achieved under which the filtering error is exponentially bounded in mean square. As a consequence, the application on the fault estimation problem is investigated, since faults resulting from external disturbances and component/actuator malfunctions might still occur in the presence of ETT and stochastic sensor saturations. The main novelty of the chapter lies in the following aspects: 1) a comprehensive model is established which covers nonlinearities, event-triggered measurement transmissions, and stochastic sensor saturations; 2) an upper bound of the filtering error covariance is minimized by appropriately designing the recursive filter and the algorithm is applied in the fault estimation problem; and 3) the boundedness of the filtering error dynamics is analyzed.

7.1 Problem Formulation

Consider the following stochastic discrete-time nonlinear system:

$$\begin{cases} x_{k+1} = g(x_k, u_k) + D_k w_k, \\ y_k = \Upsilon_{\alpha_k} \sigma(C_k x_k) + (I - \Upsilon_{\alpha_k}) C_k x_k + F_k v_k, \end{cases} \quad (7.1)$$

where $x_k \in \mathbb{R}^n$ is the state; $u_k \in \mathbb{R}^l$ is the control; $y_k \in \mathbb{R}^m$ is the measurement; $w_k \in \mathbb{R}^p$ and $v_k \in \mathbb{R}^q$ are the mutually uncorrelated zero-mean process noise and the communication noise with $\mathbb{E}\{w_k w_k^T\} = W_k$ and $\mathbb{E}\{v_k v_k^T\} = V_k$. The initial condition x_0 is stochastic with known $\mathbb{E}\{x_0\}$ and $\mathbb{E}\{x_0 x_0^T\}$, and independent of the noises. C_k, D_k, and F_k are known matrices and the nonlinear function g is twice continuously differentiable.

Problem Formulation 117

For every $k \in \mathbb{N}$, $\Upsilon_{\alpha_k} = \text{diag}\{\alpha_{1,k}, \ldots, \alpha_{m,k}\}$ where for every $i = 1, 2, \ldots, m$, $\alpha_{i,k} \in \mathbb{R}$ is a Bernoulli distributed white sequence taking values on 0 or 1 with

$$\begin{cases} \text{Prob}\{\alpha_{i,k} = 1\} = \upsilon_i, \\ \text{Prob}\{\alpha_{i,k} = 0\} = 1 - \upsilon_i. \end{cases} \quad (7.2)$$

Here, $\upsilon_i \in [0,1]$ is a known scalar for every i. Denoting $\Upsilon_\upsilon := \text{diag}\{\upsilon_1, \ldots, \upsilon_m\}$, it follows directly that $\mathbb{E}\{\Upsilon_{\alpha_k}\} = \Upsilon_\upsilon$.

For a vector $r = [r_1, \ldots, r_m]^T$, the saturation function $\sigma : \mathbb{R}^m \to \mathbb{R}^m$ is defined as:

$$\sigma(r) = [\sigma_1(r_1), \ldots, \sigma_m(r_m)]^T \quad (7.3)$$

where $\sigma_s(r_s) = \text{sign}(r_s)\min(b_s, |r_s|)$ and $b_s \geq 0$ for all $s = 1, \ldots, m$. Furthermore, $\text{sign}(\cdot)$ denotes the signum function and b_s represents the saturation level.

In this chapter, the following standard send-on-delta [117] transmission strategy is considered: the current measurement y_{k+j} would be transmitted if it satisfies

$$(y_{k+j} - y_k)^T (y_{k+j} - y_k) > \varsigma, \quad (7.4)$$

where y_k is the previously transmitted measurement and ς is a given positive scalar. Letting the release instants be denoted by $k_0, k_1, \cdots$, the released signal $\tilde{y}_k$ can be written as

$$\tilde{y}_k = y_{k_j}, k \in \{k_j, k_j + 1, \cdots, k_{j+1} - 1\}. \quad (7.5)$$

For system (7.1), consider a filter of the following structure:

$$\hat{x}_{k+1|k} = g(\hat{x}_{k|k}, u_k), \quad (7.6)$$
$$\hat{x}_{k+1|k+1} = \hat{x}_{k+1|k} + K_{k+1}\big[\tilde{y}_{k+1} - \Upsilon_\upsilon \sigma(C_{k+1}\hat{x}_{k+1|k}) - (I - \Upsilon_\upsilon)C_{k+1}\hat{x}_{k+1|k}\big], \quad (7.7)$$

where $\hat{x}_{k|k} \in \mathbb{R}^n$ is the estimation of x_k at time step k with $\hat{x}_{0|0} = \mathbb{E}\{x_0\}$, $\hat{x}_{k+1|k} \in \mathbb{R}^n$ is the one step prediction at time step k, and K_{k+1} is the filter gain to be determined.

Remark 7.1 *The measurement equation in (7.1) is introduced to describe the stochastic sensor saturations which may arise from uncertain working conditions and technological/physical limitations. The proposed transmission condition (7.4) means that the current measurement is released only when it changes greatly, i.e., the difference between the current measurement and the last transmitted one is greater than a predefined threshold in the event generator. Also note that the terms reflecting the statistics (i.e. υ_i for all $i = 1, 2, \ldots, m$) are fixed scalars, which facilitates the filter implementation. Both ETT and stochastic sensor saturations would affect the observability of the addressed system, making the filtering problem more challenging.*

118   Event-Triggered Filtering and Fault Estimation for Nonlinear Systems

Denote the prediction error, the estimation error, and their covariances conditional on the received measurements as $e_{k+1|k} = x_{k+1} - \hat{x}_{k+1|k}$, $e_{k+1|k+1} = x_{k+1} - \hat{x}_{k+1|k+1}$, $P_{k+1|k} = \mathbb{E}\left\{e_{k+1|k}e_{k+1|k}^T|y_0,\ldots,y_k\right\}$, and $P_{k+1|k+1} = \mathbb{E}\left\{e_{k+1|k+1}e_{k+1|k+1}^T|y_0,\ldots,y_{k+1}\right\}$, respectively. The goal of the addressed problem is to design an estimator in the form of (7.6) and (7.7) for system (7.1) such that an upper bound of $P_{k+1|k+1}$ can be obtained and subsequently minimized.

7.2   Filter Design

In this section, two sets of recursive Riccati-like matrix equations are established to calculate the filter parameter in (7.7) in order to minimize an upper bound of the filtering error covariance for system (7.1). To start with, it follows from (7.1) and (7.6) that

$$e_{k+1|k} = g(x_k, u_k) - g(\hat{x}_{k|k}, u_k) + D_k w_k. \tag{7.8}$$

Based on the results in [13, 178], (7.8) can be written as:

$$e_{k+1|k} = (A_k + S_k U_k)e_{k|k} + D_k w_k, \tag{7.9}$$

where

$$A_k = \left.\frac{\partial g(z_k, u_k)}{\partial z_k}\right|_{z_k = \hat{x}_{k|k}},$$

S_k is a problem-dependent scaling matrix and U_k is an unknown matrix with $\|U_k\| \leq 1$. Then, the following lemma can be established.

Lemma 7.1 *The prediction error covariance satisfies*

$$P_{k+1|k} = \mathbb{E}\left\{(A_k + S_k U_k)P_{k|k}(A_k + S_k U_k)^T\right\} + D_k W_k D_k^T, \tag{7.10}$$

and the estimation error covariance can be recursively calculated as follows:

$$\begin{aligned}
P_{k+1|k+1} =& [I - K_{k+1}(I - \Upsilon_v)C_{k+1}]P_{k+1|k}[I - K_{k+1}(I - \Upsilon_v)C_{k+1}]^T \\
&+ K_{k+1}\mathbb{E}\{(\Upsilon_{\alpha_{k+1}} - \Upsilon_v)[\sigma(C_{k+1}x_{k+1}) - C_{k+1}x_{k+1}] \\
&\times [\sigma(C_{k+1}x_{k+1}) - C_{k+1}x_{k+1}]^T(\Upsilon_{\alpha_{k+1}} - \Upsilon_v)^T\}K_{k+1}^T \\
&+ K_{k+1}\Upsilon_v\mathbb{E}\{[\sigma(C_{k+1}x_{k+1}) - \sigma(C_{k+1}\hat{x}_{k+1|k})][\sigma(C_{k+1}x_{k+1}) \\
&- \sigma(C_{k+1}\hat{x}_{k+1|k})]^T\}\Upsilon_v^T K_{k+1}^T + K_{k+1}\mathbb{E}\{(\tilde{y}_{k+1} - y_{k+1}) \\
&\times (\tilde{y}_{k+1} - y_{k+1})^T\}K_{k+1}^T - [I - K_{k+1}(I - \Upsilon_v)C_{k+1}] \\
&\times \mathbb{E}\{e_{k+1|k}[\sigma(C_{k+1}x_{k+1}) - \sigma(C_{k+1}\hat{x}_{k+1|k})]^T\}\Upsilon_v^T K_{k+1}^T
\end{aligned}$$

Filter Design

$$- K_{k+1}\Upsilon_v \mathbb{E}\{[\sigma(C_{k+1}x_{k+1}) - \sigma(C_{k+1}\hat{x}_{k+1|k})]e_{k+1|k}^T\}$$
$$[I - K_{k+1}(I - \Upsilon_v)C_{k+1}]^T - [I - K_{k+1}(I - \Upsilon_v)C_{k+1}]$$
$$\times \mathbb{E}\{e_{k+1|k}(\tilde{y}_{k+1} - y_{k+1})^T\}K_{k+1}^T - K_{k+1}\mathbb{E}\{(\tilde{y}_{k+1} - y_{k+1})$$
$$\times e_{k+1|k}^T\}[I - K_{k+1}(I - \Upsilon_v)C_{k+1}]^T + K_{k+1}\mathbb{E}\{(\tilde{y}_{k+1} - y_{k+1})$$
$$\times [\sigma(C_{k+1}x_{k+1}) - \sigma(C_{k+1}\hat{x}_{k+1|k})]^T\}\Upsilon_v^T K_{k+1}^T + K_{k+1}$$
$$\times \Upsilon_v \mathbb{E}\{[\sigma(C_{k+1}x_{k+1}) - \sigma(C_{k+1}\hat{x}_{k+1|k})](\tilde{y}_{k+1} - y_{k+1})^T\}K_{k+1}^T$$
$$+ K_{k+1}\mathbb{E}\{(\tilde{y}_{k+1} - y_{k+1})v_{k+1}^T\}F_{k+1}^T K_{k+1}^T + K_{k+1}F_{k+1}$$
$$\times \mathbb{E}\{v_{k+1}(\tilde{y}_{k+1} - y_{k+1})^T\}K_{k+1}^T + K_{k+1}\mathbb{E}\{(\Upsilon_{\alpha_{k+1}} - \Upsilon_v)$$
$$\times [\sigma(C_{k+1}x_{k+1}) - C_{k+1}x_{k+1}](\tilde{y}_{k+1} - y_{k+1})^T\}K_{k+1}^T \qquad (7.11)$$
$$+ K_{k+1}\mathbb{E}\{(\tilde{y}_{k+1} - y_{k+1})[\sigma(C_{k+1}x_{k+1}) - C_{k+1}x_{k+1}]^T$$
$$\times (\Upsilon_{\alpha_{k+1}} - \Upsilon_v)^T\}K_{k+1}^T + K_{k+1}F_{k+1}V_{k+1}F_{k+1}^T K_{k+1}^T.$$

Proof *(7.10) is easily accessible from (7.9) and the fact that $e_{k|k}$ is independent of w_k, and now we are going to prove (7.11). From (7.1) and (7.7), it follows that*

$$e_{k+1|k+1} = e_{k+1|k} - K_{k+1}[\tilde{y}_{k+1} - \Upsilon_v \sigma(C_{k+1}\hat{x}_{k+1|k}) - (I - \Upsilon_v)C_{k+1}\hat{x}_{k+1|k}]. \qquad (7.12)$$

Adding the zero term

$$K_{k+1}y_{k+1} - K_{k+1}y_{k+1} + K_{k+1}\Upsilon_v\sigma(C_{k+1}x_{k+1}) - K_{k+1}\Upsilon_v\sigma(C_{k+1}x_{k+1})$$
$$+ K_{k+1}(I - \Upsilon_v)C_{k+1}x_{k+1} - K_{k+1}(I - \Upsilon_v)C_{k+1}x_{k+1}$$

to the right-hand side of (7.12), we have

$$e_{k+1|k+1} = [I - K_{k+1}(I - \Upsilon_v)C_{k+1}]e_{k+1|k} - K_{k+1}(\Upsilon_{\alpha_{k+1}} - \Upsilon_v)[\sigma(C_{k+1}x_{k+1})$$
$$- C_{k+1}x_{k+1}] - K_{k+1}\Upsilon_v[\sigma(C_{k+1}x_{k+1}) - \sigma(C_{k+1}\hat{x}_{k+1|k})]$$
$$- K_{k+1}(\tilde{y}_{k+1} - y_{k+1}) - K_{k+1}F_{k+1}v_{k+1}. \qquad (7.13)$$

(7.11) can be obtained directly from (7.13). This concludes the proof.

Remark 7.2 *In Lemma 7.1, the exact covariances of one-step prediction error and filtering error have been obtained. However, it is very difficult to determine the covariances recursively by using these two equations because of the stochastic sensor saturations and event-triggered transmissions. To handle terms related to $y_{k+1} - \tilde{y}_{k+1}$ and the saturations, we need the posteriori PDF of the states based on the PDF of states conditional on measurements. Unfortunately, since the system (7.1) is relatively complex that contains both the nonlinearities and the stochastic sensor saturations, the conditional PDF might be difficult to calculate or approximate. In [153], $y_{k+1} - \tilde{y}_{k+1}$ is assumed to be uniformly distributed, and the filtering error covariance is updated accordingly. However, such an assumption is a bit too stringent in practice. An* alternative way is to find an upper bound of the filtering error covariance and then design the filter gain to minimize the upper bound at each time step. In this way, neither the conditional PDF nor the strong assumption on the distribution of $y_{k+1} - \tilde{y}_{k+1}$ will be required.

Before proceeding, the following lemma is to be introduced [66].

Lemma 7.2 *For any two matrices $X, Y \in \mathbb{R}^{n \times n}$, the inequality $XY^T + YX^T \leq \varepsilon XX^T + \varepsilon^{-1} YY^T$ holds where $\varepsilon > 0$ is a constant scalar.*

Now we are in a position to obtain an upper bound of the filtering error covariance and design the filter to minimize the bound.

Theorem 7.1 *Let ε_j ($j = 1, \ldots, 8$) and γ_k ($k \in \mathbb{N}$) be positive scalars. Assume that the following recursive equations*

$$\bar{P}_{k+1|k} = (1+\varepsilon_1) A_k \bar{P}_{k|k} A_k^T + \gamma_k (1+\varepsilon_1^{-1}) S_k S_k^T + D_k W_k D_k^T, \tag{7.14}$$

$$\begin{aligned}\bar{P}_{k+1|k+1} =& (1+\varepsilon_2+\varepsilon_3)\left[I - K_{k+1}(I-\Upsilon_v)C_{k+1}\right]\bar{P}_{k+1|k}\left[I - K_{k+1}(I-\Upsilon_v)C_{k+1}\right]^T \\ &+ (1+\varepsilon_6)K_{k+1}(\tilde{\Upsilon} \circ \Theta_{k+1})K_{k+1}^T + 4\bar{b}(1+\varepsilon_2^{-1}+\varepsilon_4) \\ &\times K_{k+1}\Upsilon_v \Upsilon_v^T K_{k+1}^T + \varsigma(1+\varepsilon_3^{-1}+\varepsilon_4^{-1}+\varepsilon_5+\varepsilon_6^{-1})K_{k+1}K_{k+1}^T \\ &+ (1+\varepsilon_5^{-1})K_{k+1}F_{k+1}V_{k+1}F_{k+1}^T K_{k+1}^T \end{aligned} \tag{7.15}$$

have positive definite solutions with initial condition $\bar{P}_{0|0} = P_{0|0}$, where

$$\bar{b} = \sum_{s=1}^m b_s^2, \tag{7.16}$$

$$\begin{aligned}\Theta_{k+1} =& \bar{b}(1+\varepsilon_7)I + (1+\varepsilon_7^{-1})(1+\varepsilon_8)C_{k+1}\bar{P}_{k+1|k}C_{k+1}^T \\ &+ (1+\varepsilon_7^{-1})(1+\varepsilon_8^{-1})C_{k+1}\hat{x}_{k+1|k}\hat{x}_{k+1|k}^T C_{k+1}^T,\end{aligned} \tag{7.17}$$

$$\bar{P}_{k|k} \leq \gamma_k I, \tag{7.18}$$

$$\tilde{\Upsilon} = \mathrm{diag}\{v_1 - v_1^2, \ldots, v_m - v_m^2\}, \tag{7.19}$$

$$\begin{aligned}Y_{k+1} =& (1+\varepsilon_2+\varepsilon_3)(I - \Upsilon_v)C_{k+1}\bar{P}_{k+1|k}C_{k+1}^T(I-\Upsilon_v)^T + 4\bar{b} \\ &\times (1+\varepsilon_2^{-1}+\varepsilon_4)\Upsilon_v \Upsilon_v^T + (1+\varepsilon_6)\tilde{\Upsilon} \circ \Theta_{k+1} + \varsigma(1+\varepsilon_3^{-1} \\ &+ \varepsilon_4^{-1}+\varepsilon_5+\varepsilon_6^{-1})I + (1+\varepsilon_5^{-1})F_{k+1}V_{k+1}F_{k+1}^T,\end{aligned} \tag{7.20}$$

$$Z_{k+1} = (1+\varepsilon_2+\varepsilon_3)(I-\Upsilon_v)C_{k+1}\bar{P}_{k+1|k}, \tag{7.21}$$

$$K_{k+1} = Z_{k+1}^T Y_{k+1}^{-1}. \tag{7.22}$$

Then, $\bar{P}_{k|k}$ is an upper bound of $P_{k|k}$, and the bound $\bar{P}_{k+1|k+1}$ is minimized at each time step with the filter gain K_{k+1} given in (7.22).

Proof The theorem can be proved by induction. Based on the initial condition, we have $\bar{P}_{0|0} \geq P_{0|0}$. Then, assume that $\bar{P}_{k|k} \geq P_{k|k}$, and we

Filter Design

need to prove that $\bar{P}_{k+1|k+1} \geq P_{k+1|k+1}$. Firstly, based on $\bar{P}_{k|k} \geq P_{k|k}$, one needs to show that $\bar{P}_{k+1|k} \geq P_{k+1|k}$ and $\Theta_{k+1} \geq \mathbb{E}\{[\sigma(C_{k+1}x_{k+1}) - C_{k+1}x_{k+1}][\sigma(C_{k+1}x_{k+1}) - C_{k+1}x_{k+1}]^T\} =: \Psi_{k+1}$.

With the assumption $\bar{P}_{k|k} \geq P_{k|k}$, we have from (7.10) that

$$P_{k+1|k} \leq \mathbb{E}\left\{(A_k + S_k U_k)\bar{P}_{k|k}(A_k + S_k U_k)^T\right\} + D_k W_k D_k^T.$$

Then, it follows from Lemma 2 that

$$P_{k+1|k} \leq (1+\varepsilon_1)A_k \bar{P}_{k|k} A_k^T + (1+\varepsilon_1^{-1})\mathbb{E}\left\{S_k U_k \bar{P}_{k|k} U_k^T S_k^T\right\} + D_k W_k D_k^T. \tag{7.23}$$

From (7.18) and $\|U_k\| \leq 1$, we have $S_k U_k \bar{P}_{k|k} U_k^T S_k^T \leq \gamma_k S_k S_k^T$ and, subsequently, (7.23) can be written as

$$P_{k+1|k} \leq (1+\varepsilon_1)A_k \bar{P}_{k|k} A_k^T + \gamma_k(1+\varepsilon_1^{-1})S_k S_k^T + D_k W_k D_k^T = \bar{P}_{k+1|k}.$$

Next, let us deal with Ψ_{k+1}. It follows from Lemma 7.2 that

$$\Psi_{k+1} \leq (1+\varepsilon_7)\mathbb{E}\{\sigma(C_{k+1}x_{k+1})\sigma^T(C_{k+1}x_{k+1})\} + (1+\varepsilon_7^{-1})$$
$$\times \mathbb{E}\{C_{k+1}x_{k+1}x_{k+1}^T C_{k+1}^T\}.$$

From the facts that $x_{k+1} = \hat{x}_{k+1|k} + e_{k+1|k}$ and $\bar{P}_{k+1|k} \geq P_{k+1|k}$, it follows that

$$\Psi_{k+1} \leq (1+\varepsilon_7)\mathbb{E}\{\sigma(C_{k+1}x_{k+1})\sigma^T(C_{k+1}x_{k+1})\} + (1+\varepsilon_7^{-1})(1+\varepsilon_8)C_{k+1}$$
$$\times \bar{P}_{k+1|k} C_{k+1}^T + (1+\varepsilon_7^{-1})(1+\varepsilon_8^{-1})C_{k+1}\hat{x}_{k+1|k}\hat{x}_{k+1|k}^T C_{k+1}^T. \tag{7.24}$$

Since the absolute value of the ith entry of $\sigma(C_{k+1}x_{k+1})$ is less than or equal to b_i, we obtain

$$\sigma(C_{k+1}x_{k+1})\sigma^T(C_{k+1}x_{k+1}) \leq \bar{b}I. \tag{7.25}$$

Substituting (7.25) into (7.24) yields

$$\Psi_{k+1} \leq \bar{b}(1+\varepsilon_7)I + (1+\varepsilon_7^{-1})(1+\varepsilon_8)C_{k+1}\bar{P}_{k+1|k}C_{k+1}^T$$
$$+ (1+\varepsilon_7^{-1})(1+\varepsilon_8^{-1})C_{k+1}\hat{x}_{k+1|k}\hat{x}_{k+1|k}^T C_{k+1}^T = \Theta_{k+1}.$$

Now, we are going to show that $\bar{P}_{k+1|k+1} \geq P_{k+1|k+1}$. It follows from Lemma 2 that

$$P_{k+1|k+1} \leq (1+\varepsilon_2+\varepsilon_3)\left[I - K_{k+1}(I-\Upsilon_v)C_{k+1}\right]P_{k+1|k}[I - K_{k+1}(I-\Upsilon_v)$$
$$\times C_{k+1}]^T + (1+\varepsilon_6)K_{k+1}\mathbb{E}\{(\Upsilon_{\alpha_{k+1}} - \Upsilon_v)[\sigma(C_{k+1}x_{k+1}) - C_{k+1}x_{k+1}]$$
$$\times [\sigma(C_{k+1}x_{k+1}) - C_{k+1}x_{k+1}]^T(\Upsilon_{\alpha_{k+1}} - \Upsilon_v)^T\}K_{k+1}^T$$
$$+ (1+\varepsilon_2^{-1}+\varepsilon_4)K_{k+1}\Upsilon_v\mathbb{E}\{[\sigma(C_{k+1}x_{k+1}) - \sigma(C_{k+1}\hat{x}_{k+1|k})]$$
$$\times [\sigma(C_{k+1}x_{k+1}) - \sigma(C_{k+1}\hat{x}_{k+1|k})]^T\}\Upsilon_v^T K_{k+1}^T$$
$$+ (1+\varepsilon_3^{-1}+\varepsilon_4^{-1}+\varepsilon_5+\varepsilon_6^{-1})K_{k+1}\mathbb{E}\{(\tilde{y}_{k+1} - y_{k+1})(\tilde{y}_{k+1} - y_{k+1})^T\}K_{k+1}^T$$
$$+ (1+\varepsilon_5^{-1})K_{k+1}F_{k+1}V_{k+1}F_{k+1}^T K_{k+1}^T. \tag{7.26}$$

Considering $\bar{P}_{k+1|k} \geq P_{k+1|k}$ and $\Theta_{k+1} \geq \Psi_{k+1}$, (7.26) can be written as

$$\begin{aligned}P_{k+1|k+1} \leq &(1+\varepsilon_2+\varepsilon_3)[I-K_{k+1}(I-\Upsilon_v)C_{k+1}]\bar{P}_{k+1|k}[I-K_{k+1}(I-\Upsilon_v)\\&\times C_{k+1}]^T + (1+\varepsilon_6)K_{k+1}\mathbb{E}\{(\Upsilon_{\alpha_{k+1}}-\Upsilon_v)\Theta_{k+1}(\Upsilon_{\alpha_{k+1}}-\Upsilon_v)^T\}\\&\times K_{k+1}^T + (1+\varepsilon_2^{-1}+\varepsilon_4)K_{k+1}\Upsilon_v\mathbb{E}\{[\sigma(C_{k+1}x_{k+1})\\&-\sigma(C_{k+1}\hat{x}_{k+1|k})][\sigma(C_{k+1}x_{k+1})-\sigma(C_{k+1}\hat{x}_{k+1|k})]^T\}\\&\times \Upsilon_v^T K_{k+1}^T + (\varepsilon_3^{-1}+\varepsilon_4^{-1}+\varepsilon_5+\varepsilon_6^{-1}+1)K_{k+1}\\&\times \mathbb{E}\{(\tilde{y}_{k+1}-y_{k+1})(\tilde{y}_{k+1}-y_{k+1})^T\}K_{k+1}^T\\&+ (1+\varepsilon_5^{-1})K_{k+1}F_{k+1}V_{k+1}F_{k+1}^T K_{k+1}^T.\end{aligned}$$

Based on the transmission condition (7.4), for any $k \in \mathbb{N}$, we have

$$(\tilde{y}_k - y_k)(\tilde{y}_k - y_k)^T \leq \varsigma I. \tag{7.27}$$

Similar to (7.25), we get

$$[\sigma(C_{k+1}x_{k+1}) - \sigma(C_{k+1}\hat{x}_{k+1|k})][\sigma(C_{k+1}x_{k+1}) - \sigma(C_{k+1}\hat{x}_{k+1|k})]^T \leq 4\bar{b}I. \tag{7.28}$$

From (7.27) and (7.28), it follows that

$$\begin{aligned}P_{k+1|k+1} \leq &(1+\varepsilon_2+\varepsilon_3)\left[I-K_{k+1}(I-\Upsilon_v)C_{k+1}\right]\bar{P}_{k+1|k}[I-K_{k+1}(I-\Upsilon_v)\\&\times C_{k+1}]^T + (1+\varepsilon_6)K_{k+1}(\tilde{\Upsilon}\circ\Theta_{k+1})K_{k+1}^T + 4\bar{b}(1+\varepsilon_2^{-1}+\varepsilon_4)\\&\times K_{k+1}\Upsilon_v\Upsilon_v^T K_{k+1}^T + \varsigma(1+\varepsilon_3^{-1}+\varepsilon_4^{-1}+\varepsilon_5+\varepsilon_6^{-1})K_{k+1}K_{k+1}^T\\&+ (1+\varepsilon_5^{-1})K_{k+1}F_{k+1}V_{k+1}F_{k+1}^T K_{k+1}^T = \bar{P}_{k+1|k+1}.\end{aligned}$$

So far, $\bar{P}_{k|k}$ has been verified to be an upper bound of $P_{k|k}$, and what remains to show is that K_{k+1} in (7.22) minimizes the bound. With (7.20) and (7.21), $\bar{P}_{k+1|k+1}$ can be written as

$$\bar{P}_{k+1|k+1} = (1+\varepsilon_2+\varepsilon_3)\bar{P}_{k+1|k} + K_{k+1}Y_{k+1}K_{k+1}^T - Z_{k+1}^T K_{k+1}^T - K_{k+1}Z_{k+1}.$$

Noticing the fact that $Y_{k+1} = Y_{k+1}^T > 0$ and completing the square with respect to K_{k+1}, we have

$$\begin{aligned}\bar{P}_{k+1|k+1} =& (K_{k+1} - Z_{k+1}^T Y_{k+1}^{-1})Y_{k+1}(K_{k+1} - Z_{k+1}^T Y_{k+1}^{-1})^T \\&- Z_{k+1}^T Y_{k+1}^{-1} Z_{k+1} + (1+\varepsilon_2+\varepsilon_3)\bar{P}_{k+1|k}.\end{aligned}$$

Therefore, it is straightforward to see that when $K_{k+1} = Z_{k+1}^T Y_{k+1}^{-1}$, the bound $\bar{P}_{k+1|k+1}$ is minimized and satisfies the next recursion:

$$\bar{P}_{k+1|k+1} = -Z_{k+1}^T Y_{k+1}^{-1} Z_{k+1} + (1+\varepsilon_2+\varepsilon_3)\bar{P}_{k+1|k}.$$

The proof is now complete.

Boundedness Analysis 123

Remark 7.3 *The filtering problem has been solved in Theorem 7.1 in a recursive way for a class of discrete time-varying nonlinear systems with stochastic sensor saturations and event-triggered transmissions. To cope with the stochastic sensor saturations and event-triggered transmissions, special effort has been made to obtain the upper bound of the filtering error covariance and design the filter so as to minimize the bound. The matrix S_k reflects the linearization errors, the parameters Υ_v and $\bar{b}$ represent the effects of stochastic sensor saturations, and the scalar ς quantities the influences of the event-triggered transmissions. The parameters ε_j can be determined to balance the intrinsic characteristic of the proposed filter and the impacts induced by ETT and stochastic sensor saturations. Neither the approximated PDF of states conditional on measurements nor the assumption on the distribution of $y_{k+1} - \tilde{y}_{k+1}$ is required in the presented approach. In other words, the applicability and feasibility of the algorithm have been enhanced. Furthermore, the desired filter gain is obtained via solving two sets of discrete Riccati-like equations, hence the method is suitable for online applications.*

7.3 Boundedness Analysis

Before proceeding, the following widely used concept for the boundedness of stochastic processes is introduced.

Definition 7.1 *[136] The stochastic process ζ_k is said to be exponentially bounded in mean square if there are real numbers $\eta > 0$, $\nu > 0$ and $0 < \vartheta < 1$ such that*

$$\mathbb{E}\left\{\|\zeta_k\|^2\right\} \leq \eta \|\zeta_0\|^2 \vartheta^k + \nu \qquad (7.29)$$

holds for every $k > 0$.

For the boundedness analysis of the estimation error, we establish sufficient conditions under which the filtering error is exponentially bounded in mean square. For this purpose, we make the following assumption.

Assumption 7.1 *There are positive real numbers $\bar{a}, \bar{c}, \underline{c}, \bar{v}, \underline{v}, \bar{\psi}, \bar{s}, \bar{f}, \underline{f}, \bar{d}, \underline{d}, \bar{w}, \underline{w}, \bar{v} > 0$ such that the following bounds on various matrices are fulfilled for every $1 \leq i \leq m$ and $k \geq 0$:*

$$\|A_k\| \leq \bar{a}, \ \|S_k\| \leq \bar{s}, \ \underline{c} \leq \|C_k\| \leq \bar{c}, \mathrm{tr}\{\Psi_k\} \leq \bar{\psi}, \ \underline{v} \leq v_i \leq \bar{v},$$
$$\underline{d}I \leq D_k D_k^T \leq \bar{d}I, \ \underline{f} \leq \|F_k\| \leq \bar{f}, \underline{w}I \leq W_k \leq \bar{w}I, \ V_k \leq \bar{v}I. \qquad (7.30)$$

Moreover, the following inequality holds:

$$\varrho = \left[(1+\eta)\bar{a}^2 + (1+\eta^{-1})\bar{s}^2\right]\left[1 + \frac{\bar{c}^2}{(1-\bar{v})\underline{c}^2}\right]^2 < 1, \qquad (7.31)$$

where η is a positive scalar.

Theorem 7.2 *Consider the time-varying system (7.1) with the filter given in (7.6) and (7.7) whose parameters are provided in Theorem 7.1. Under Assumption 7.1, the filtering error is exponentially bounded in mean square.*

Proof Denote $\Xi_{k+1} = I - K_{k+1}(I - \Upsilon_v)C_{k+1}$, $\check{A}_{k+1} = \Xi_{k+1}A_k$ and $\hat{A}_{k+1} = \Xi_{k+1}S_kU_k$. Substituting (7.9) into (7.13) and considering the definitions of $\check{A}_{k+1}$ and $\hat{A}_{k+1}$, we have

$$e_{k+1|k+1} = (\check{A}_{k+1} + \hat{A}_{k+1})e_{k|k} + p_{k+1} + q_{k+1}, \tag{7.32}$$

where

$$p_{k+1} = -K_{k+1}\Upsilon_v[\sigma(C_{k+1}x_{k+1}) - \sigma(C_{k+1}\hat{x}_{k+1|k})] - K_{k+1}(\tilde{y}_{k+1} - y_{k+1}),$$
$$q_{k+1} = \Xi_{k+1}D_kw_k - K_{k+1}(\Upsilon_{\alpha_{k+1}} - \Upsilon_v)[\sigma(C_{k+1}x_{k+1}) - C_{k+1}x_{k+1}]$$
$$\quad - K_{k+1}F_{k+1}v_{k+1}.$$

Based on (7.22), it follows easily from $\underline{v} \le v_i \le \bar{v}$ and $\underline{c} \le \|C_{k+1}\| \le \bar{c}$ that

$$\|K_{k+1}\| = \|Z_{k+1}^T Y_{k+1}^{-1}\|$$
$$< \left\| [(1+\varepsilon_2+\varepsilon_3)(I-\Upsilon_v)C_{k+1}\bar{P}_{k+1|k}]^T [(1+\varepsilon_2+\varepsilon_3) \right.$$
$$\left. \times (I-\Upsilon_v)C_{k+1}\bar{P}_{k+1|k}C_{k+1}^T(I-\Upsilon_v)^T]^{-1} \right\|$$
$$\le \frac{\bar{c}}{(1-\bar{v})\underline{c}^2} =: \bar{k},$$

and

$$\|\Xi_{k+1}\| < \left\| I - [(1+\varepsilon_2+\varepsilon_3)(I-\Upsilon_v)C_{k+1}\bar{P}_{k+1|k}]^T [(1+\varepsilon_2+\varepsilon_3)(I-\Upsilon_v) \right.$$
$$\left. \times C_{k+1}\bar{P}_{k+1|k}C_{k+1}^T(I-\Upsilon_v)^T]^{-1}(I-\Upsilon_v)C_{k+1} \right\|$$
$$\le 1 + \frac{\bar{c}^2}{\underline{c}^2} =: \bar{\xi}.$$

Then, we have

$$\|\check{A}_{k+1}\| \le \|\Xi_{k+1}\|\|A_k\| \le \bar{\xi}\bar{a} =: \bar{a}_1,$$
$$\|\hat{A}_{k+1}\| \le \|\Xi_{k+1}\|\|S_k\|\|U_k\| \le \bar{\xi}\bar{s} =: \bar{a}_2.$$

Recalling Lemma 7.2, we can obtain

$$\mathbb{E}\left\{p_{k+1}^T p_{k+1}\right\}$$
$$\le (1+\eta_1)\mathbb{E}\left\{[\sigma(C_{k+1}x_{k+1}) - \sigma(C_{k+1}\hat{x}_{k+1|k})]^T \Upsilon_v^T K_{k+1}^T K_{k+1}\Upsilon_v[\sigma(C_{k+1}x_{k+1}) \right.$$
$$\left. - \sigma(C_{k+1}\hat{x}_{k+1|k})]\right\} + (1+\eta_1^{-1})\mathbb{E}\left\{(\tilde{y}_{k+1} - y_{k+1})^T K_{k+1}^T K_{k+1}(\tilde{y}_{k+1} - y_{k+1})\right\},$$
$$\tag{7.33}$$

Boundedness Analysis

where η_1 is a positive scalar.

Substituting (7.27) and (7.28) into (7.33) leads to

$$\mathbb{E}\left\{p_{k+1}^T p_{k+1}\right\} \leq 4(1+\eta_1)\bar{b}\bar{v}^2\bar{k}^2 + (1+\eta_1^{-1})\varsigma\bar{k}^2 =: \bar{p}^2.$$

Since w_k, v_k, and $\alpha_{i,k}$ are assumed to be mutually independent, we have

$$\mathbb{E}\left\{q_{k+1}^T q_{k+1}\right\}$$
$$=\mathbb{E}\left\{w_k^T D_k^T \Xi_{k+1}^T \Xi_{k+1} D_k w_k\right\} + \mathbb{E}\left\{v_{k+1}^T F_{k+1}^T K_{k+1}^T K_{k+1} F_{k+1} v_{k+1}\right\}$$
$$+ \mathbb{E}\big\{[\sigma(C_{k+1}x_{k+1}) - C_{k+1}x_{k+1}]^T (\Upsilon_{\alpha_{k+1}} - \Upsilon_v)^T K_{k+1}^T K_{k+1} (\Upsilon_{\alpha_{k+1}} - \Upsilon_v)$$
$$\times [\sigma(C_{k+1}x_{k+1}) - C_{k+1}x_{k+1}]\big\}$$
$$\leq p\bar{\xi}^2 \bar{d}^2 \bar{w} + q\bar{k}^2 \bar{f}^2 \bar{v} + \bar{k}^2 \hat{v}^2 \bar{\psi} =: \bar{q}^2,$$

where $\hat{v} = \max\{1 - \underline{v}, \bar{v}\}$.

Consider the following iterative matrix equation

$$\Pi_{k+1} = (1+\eta)\check{A}_{k+1}\Pi_k\check{A}_{k+1}^T + (1+\eta^{-1})\rho_{\max}(\Pi_k)\Xi_{k+1}S_kS_k^T\Xi_{k+1}^T + D_kW_kD_k^T,$$

with initial condition $\Pi_0 = D_0W_0D_0^T$ where $\rho_{\max}(\Pi_k)$ represents the maximum eigenvalue of Π_k. Then, it follows directly that

$$\|\Pi_{k+1}\| \leq \|\Pi_k\|\left[(1+\eta)\|\check{A}_{k+1}\|^2 + (1+\eta^{-1})\|\Xi_{k+1}S_k\|^2\right] + \|D_kW_kD_k^T\|$$
$$\leq \varrho\|\Pi_k\| + \bar{w}\bar{d}^2.$$

By iteration, we obtain

$$\|\Pi_k\| \leq \varrho^k\|\Pi_0\| + \bar{w}\bar{d}^2\sum_{i=0}^{k-1}\varrho^i.$$

With assumption (7.31), we have $\varrho < 1$ and then arrive at

$$\|\Pi_k\| < \|\Pi_0\| + \bar{w}\bar{d}^2\sum_{i=0}^{\infty}\varrho^i = \|\Pi_0\| + \frac{\bar{w}\bar{d}^2}{1-\varrho}. \tag{7.34}$$

Furthermore, since Π_k is positive definite for all k, it is straightforward to see that

$$\Pi_{k+1} \geq D_kW_kD_k^T \geq \underline{w}\underline{d}^2 I. \tag{7.35}$$

Based on (7.34) and (7.35), it can be concluded that there are positive real numbers $\underline{\pi}, \bar{\pi} > 0$ such that the inequality $\underline{\pi}I \leq \Pi_k \leq \bar{\pi}I$ holds for every $k \geq 0$.

According to Assumption 7.1, we have

$$(\check{A}_{k+1} + \hat{A}_{k+1})^T \Pi_{k+1}^{-1}(\check{A}_{k+1} + \hat{A}_{k+1}) - \Pi_k^{-1}$$

$$\leq (\check{A}_{k+1} + \hat{A}_{k+1})^T \left[(\check{A}_{k+1} + \hat{A}_{k+1})\Pi_k(\check{A}_{k+1} + \hat{A}_{k+1})^T + D_k W_k D_k^T \right]^{-1}$$

$$(\check{A}_{k+1} + \hat{A}_{k+1}) - \Pi_k^{-1}$$

$$= -\left[\Pi_k + \Pi_k(\check{A}_{k+1} + \hat{A}_{k+1})^T \left(D_k W_k D_k^T \right)^{-1} (\check{A}_{k+1} + \hat{A}_{k+1})\Pi_k \right]^{-1}$$

$$= -\left[I + (\check{A}_{k+1} + \hat{A}_{k+1})^T \left(D_k W_k D_k^T \right)^{-1} (\check{A}_{k+1} + \hat{A}_{k+1})\Pi_k \right]^{-1} \Pi_k^{-1}$$

$$\leq -\left[\frac{(\bar{a}_1 + \bar{a}_2)^2 \bar{\pi}}{\underline{w}\underline{d}^2} + 1 \right]^{-1} \Pi_k^{-1}.$$

Define $\alpha_0 = \left[\frac{(\bar{a}_1 + \bar{a}_2)^2 \bar{\pi}}{\underline{w}\underline{d}^2} + 1 \right]^{-1}$. Since $\alpha_0 < 1$, there always exists a positive scalar β such that $\alpha = (1 + \beta)(1 - \alpha_0) < 1$. Choosing a positive scalar $\gamma > 0$, and denoting $V_k(e_{k|k}) = e_{k|k}^T \Pi_k^{-1} e_{k|k}$, and $\mu = \left[(1 + \beta^{-1} + \gamma)\bar{p}^2 + (1 + \gamma^{-1})\bar{q}^2 \right] / \underline{\pi}$, it follows from (7.32) that

$$\mathbb{E}\left\{ V_{k+1}(e_{k+1|k+1}) | e_{k|k} \right\} - (1 + \beta)V_k(e_{k|k})$$

$$= \mathbb{E}\left\{ \left[(\check{A}_{k+1} + \hat{A}_{k+1})e_{k|k} + p_{k+1} + q_{k+1} \right]^T \right.$$

$$\left. \times \Pi_{k+1}^{-1} \left[(\check{A}_{k+1} + \hat{A}_{k+1})e_{k|k} + p_{k+1} + q_{k+1} \right] | e_{k|k} \right\} - (1 + \beta)e_{k|k}^T \Pi_k^{-1} e_{k|k}$$

$$= \mathbb{E}\left\{ e_{k|k}^T (\check{A}_{k+1} + \hat{A}_{k+1})^T \Pi_{k+1}^{-1} (\check{A}_{k+1} + \hat{A}_{k+1})e_{k|k} - (1 + \beta)e_{k|k}^T \Pi_k^{-1} e_{k|k} | e_{k|k} \right\}$$

$$+ 2\mathbb{E}\left\{ e_{k|k}^T (\check{A}_{k+1} + \hat{A}_{k+1})^T \Pi_{k+1}^{-1} p_{k+1} | e_{k|k} \right\} + \mathbb{E}\left\{ p_{k+1}^T \Pi_{k+1}^{-1} p_{k+1} | e_{k|k} \right\}$$

$$+ 2\mathbb{E}\left\{ p_{k+1}^T \Pi_{k+1}^{-1} q_{k+1} | e_{k|k} \right\} + \mathbb{E}\left\{ q_{k+1}^T \Pi_{k+1}^{-1} q_{k+1} | e_{k|k} \right\}. \tag{7.36}$$

Applying Lemma 7.2 to (7.36), we have

$$\mathbb{E}\left\{ V_{k+1}(e_{k+1|k+1}) | e_{k|k} \right\} - (1 + \beta)V_k(e_{k|k})$$

$$\leq (1 + \beta)\mathbb{E}\left\{ e_{k|k}^T \left[(\check{A}_{k+1} + \hat{A}_{k+1})^T \Pi_{k+1}^{-1} (\check{A}_{k+1} + \hat{A}_{k+1}) - \Pi_k^{-1} \right] e_{k|k} | e_{k|k} \right\}$$

$$+ (1 + \beta^{-1} + \gamma)\mathbb{E}\left\{ p_{k+1}^T \Pi_{k+1}^{-1} p_{k+1} | e_{k|k} \right\} + (1 + \gamma^{-1})\mathbb{E}\left\{ q_{k+1}^T \Pi_{k+1}^{-1} q_{k+1} | e_{k|k} \right\}$$

$$\leq -\alpha_0(1 + \beta)V_k(e_{k|k}) + \mu.$$

Then, it follows that

$$\mathbb{E}\left\{ V_{k+1}(e_{k+1|k+1}) | e_{k|k} \right\} \leq \alpha V_k(e_{k|k}) + \mu,$$

which gives rise to

$$\mathbb{E}\left\{ \|e_{k|k}\|^2 \right\} \leq \frac{\bar{\pi}}{\underline{\pi}} \|e_{0|0}\|^2 \alpha^k + \mu \bar{\pi} \sum_{i=0}^{\infty} \alpha^i$$

$$= \frac{\bar{\pi}}{\underline{\pi}} \|e_{0|0}\|^2 \alpha^k + \frac{\mu \bar{\pi}}{1 - \alpha}$$

Fault Estimation 127

in which the relationships $0 < \alpha < 1$ *and* $\mu, \bar{\pi} > 0$ *have been utilized. Therefore, the stochastic process* $e_{k|k}$ *is exponentially bounded in mean square and the proof is complete.*

7.4 Fault Estimation

In this section, we aim to show that the main results in Theorem 7.1 can be applied to estimate both the system state and additive faults within a unified framework.

Consider the following faulty system corresponding to (7.1):

$$\begin{cases} x_{k+1} = g(x_k, u_k) + D_k w_k + E_k f_k, \\ y_k = \Upsilon_{\alpha_k} \sigma(C_k x_k) + (I - \Upsilon_{\alpha_k}) C_k x_k + F_k v_k, \end{cases} \quad (7.37)$$

where $f_k \in \mathbb{R}^l$ is the additive fault, E_k is a known matrix with appropriate dimensions, and all the other variables are the same as defined in (7.1). Defining an augmented state $\bar{x}_k = \begin{bmatrix} x_k^T, f_k^T \end{bmatrix}^T$, (7.37) can be rewritten as follows:

$$\begin{cases} \bar{x}_{k+1} = \bar{g}(\bar{x}_k, u_k) + \bar{D}_k w_k, \\ y_k = \Upsilon_{\alpha_k} \sigma(\bar{C}_k \bar{x}_k) + (I - \Upsilon_{\alpha_k}) \bar{C}_k \bar{x}_k + F_k v_k, \end{cases} \quad (7.38)$$

where

$$\bar{g}(\bar{x}_k, u_k) := \begin{bmatrix} g(x_k, u_k) + E_k f_k \\ f_k \end{bmatrix}, \; \bar{D}_k := \begin{bmatrix} D_k \\ 0 \end{bmatrix}, \; \bar{C}_k := [C_k, 0].$$

Similar to (7.6) and (7.7), consider a filter of the following structure:

$$\tilde{x}_{k+1|k} = \bar{g}(\tilde{x}_{k|k}, u_k), \quad (7.39)$$

$$\tilde{x}_{k+1|k+1} = \tilde{x}_{k+1|k} + \tilde{K}_{k+1} \big[\tilde{y}_{k+1} - \Upsilon_v \sigma(\bar{C}_{k+1} \tilde{x}_{k+1|k}) - (I - \Upsilon_v) \bar{C}_{k+1} \tilde{x}_{k+1|k} \big], \quad (7.40)$$

where $\tilde{x}_{k|k} \in \mathbb{R}^n$ is the estimation of $\bar{x}_k$ at time step k with $\tilde{x}_{0|0} = \begin{bmatrix} \mathbb{E}\{x_0^T\}, 0^T \end{bmatrix}^T$, $\tilde{x}_{k+1|k} \in \mathbb{R}^n$ is the one step prediction at time step k, and $\tilde{K}_{k+1}$ is the filter gain to be determined. Denote the prediction error, the estimation error, and their covariances conditional on the received measurements as $\tilde{e}_{k+1|k} = \bar{x}_{k+1} - \tilde{x}_{k+1|k}$, $\tilde{e}_{k+1|k+1} = \bar{x}_{k+1} - \tilde{x}_{k+1|k+1}$, $Q_{k+1|k} = \mathbb{E}\{\tilde{e}_{k+1|k} \tilde{e}_{k+1|k}^T | y_0, \ldots, y_k\}$, and $Q_{k+1|k+1} = \mathbb{E}\{\tilde{e}_{k+1|k+1} \tilde{e}_{k+1|k+1}^T | y_0, \ldots, y_{k+1}\}$, respectively. Then, we can obtain the following theorem whose proof is similar with that of Theorem 7.1 and is therefore omitted here.

128 Event-Triggered Filtering and Fault Estimation for Nonlinear Systems

Theorem 7.3 *Let $\tilde{\varepsilon}_j (j = 1, \ldots, 8)$ and $\tilde{\gamma}_k (k \in \mathbb{N})$ be positive scalars. Assume that, with initial condition $\bar{Q}_{0|0} = Q_{0|0}$, the following equations*

$$\bar{Q}_{k+1|k} = (1 + \tilde{\varepsilon}_1)\tilde{A}_k \bar{Q}_{k|k} \tilde{A}_k^T + \tilde{\gamma}_k(1 + \tilde{\varepsilon}_1^{-1})\tilde{S}_k \tilde{S}_k^T + \bar{D}_k W_k \bar{D}_k^T, \tag{7.41}$$

$$\begin{aligned}\bar{Q}_{k+1|k+1} =& (1 + \tilde{\varepsilon}_2 + \tilde{\varepsilon}_3)\left[I - \tilde{K}_{k+1}(I - \Upsilon_v)\bar{C}_{k+1}\right]\bar{Q}_{k+1|k}\left[I - \tilde{K}_{k+1}(I - \Upsilon_v)\right.\\ &\times \left.\bar{C}_{k+1}\right]^T + (1 + \tilde{\varepsilon}_6)\tilde{K}_{k+1}(\tilde{\Upsilon} \circ \tilde{\Theta}_{k+1})\tilde{K}_{k+1}^T + 4\bar{b}(1 + \tilde{\varepsilon}_2^{-1} + \tilde{\varepsilon}_4)\\ &\times \tilde{K}_{k+1}\Upsilon_v \Upsilon_v^T \tilde{K}_{k+1}^T + \varsigma(1 + \tilde{\varepsilon}_3^{-1} + \tilde{\varepsilon}_4^{-1} + \tilde{\varepsilon}_5 \tilde{\varepsilon}_6^{-1})\tilde{K}_{k+1}\tilde{K}_{k+1}^T\\ &+ (1 + \tilde{\varepsilon}_5^{-1})\tilde{K}_{k+1}F_{k+1}V_{k+1}F_{k+1}^T\tilde{K}_{k+1}^T, \end{aligned}\tag{7.42}$$

have positive definite solutions, where

$$\tilde{A}_k = \left.\frac{\partial \bar{g}(\bar{z}_k, u_k)}{\partial \bar{z}_k}\right|_{\bar{z}_k = \tilde{x}_{k|k}} = \begin{bmatrix} \left.\frac{\partial g(z_k, u_k)}{\partial z_k}\right|_{z_k = H\tilde{x}_{k|k}} & E_k \\ 0 & I \end{bmatrix}, \tag{7.43}$$

$$H = [I \; 0], \tag{7.44}$$

$$\begin{aligned}\tilde{\Theta}_{k+1} =& \bar{b}(1 + \tilde{\varepsilon}_7)I + (1 + \tilde{\varepsilon}_7^{-1})(1 + \tilde{\varepsilon}_8)\bar{C}_{k+1}\bar{Q}_{k+1|k}\bar{C}_{k+1}^T\\ &+ (1 + \tilde{\varepsilon}_7^{-1})(1 + \tilde{\varepsilon}_8^{-1})\bar{C}_{k+1}\tilde{x}_{k+1|k}\tilde{x}_{k+1|k}^T\bar{C}_{k+1}^T, \end{aligned}\tag{7.45}$$

$$\bar{Q}_{k|k} \leq \tilde{\gamma}_k I, \tag{7.46}$$

$$\begin{aligned}\tilde{Y}_{k+1} =& (1 + \tilde{\varepsilon}_2 + \tilde{\varepsilon}_3)(I - \Upsilon_v)\bar{C}_{k+1}\bar{Q}_{k+1|k}\bar{C}_{k+1}^T(I - \Upsilon_v)^T + 4\bar{b}\\ &\times (1 + \tilde{\varepsilon}_2^{-1} + \tilde{\varepsilon}_4)\Upsilon_v \Upsilon_v^T + \tilde{\Upsilon} \circ \tilde{\Theta}_{k+1} + \varsigma(1 + \tilde{\varepsilon}_3^{-1} + \tilde{\varepsilon}_4^{-1}\\ &+ \tilde{\varepsilon}_5)I + (1 + \tilde{\varepsilon}_5^{-1})F_{k+1}V_{k+1}F_{k+1}^T,\end{aligned}\tag{7.47}$$

$$\tilde{Z}_{k+1} = (1 + \tilde{\varepsilon}_2 + \tilde{\varepsilon}_3)(I - \Upsilon_v)\bar{C}_{k+1}\bar{Q}_{k+1|k}, \tag{7.48}$$

$$\tilde{K}_{k+1} = \tilde{Z}_{k+1}^T \tilde{Y}_{k+1}^{-1}. \tag{7.49}$$

$\tilde{S}_k$ *is a problem-dependent scaling matrix. $\bar{b}$ and $\tilde{\Upsilon}$ are the same as defined in (7.16) and (7.19), respectively. Then, $\bar{Q}_{k|k}$ is an upper bound of $Q_{k|k}$, and the bound $\bar{Q}_{k|k}$ is minimized at each time step with the filter gain given in (7.49).*

The proof is similar with that of Theorem 7.1 and is therefore omitted here. With Theorem 7.3, the state estimation and fault diagnosis problems can get solved simultaneously. The possible fault and state have been regarded as an augmented state and jointly estimated in the proposed filter. In the next section, a simulation example is illustrated to show the effectiveness of the proposed filter.

7.5 Illustrations

Inspired by the model proposed in [94], the following inverted pendulum example is considered in this section:

$$\begin{bmatrix} x_{k+1}^{(1)} \\ x_{k+1}^{(2)} \end{bmatrix} = \begin{bmatrix} 1 & \dfrac{T}{ml^2} \\ -\kappa T + Tk_1 & 1 - \dfrac{T\chi}{ml^2} + Tk_2 \end{bmatrix} \begin{bmatrix} x_k^{(1)} \\ x_k^{(2)} \end{bmatrix} + \begin{bmatrix} 0 \\ Tmgl\sin(x_k^{(1)}) \end{bmatrix}$$
$$+ \begin{bmatrix} 0 \\ 2T \end{bmatrix} w_k.$$

where $x_1 = \theta$, $x_2 = ml^2\dot{\theta}$, m is the mass, l is the length of the inverted pendulum, T is the sampling period, g is the gravitation coefficient, θ is the inclination angle, χ is the spring coefficient, κ is the damping parameter. The output measurement with stochastic sensor saturation can be written as

$$y_k = \upsilon_k \sigma\left(0.1x_k^{(1)} + 0.1x_k^{(2)}\right) + (1 - \upsilon_k)\left(0.1x_k^{(1)} + 0.1x_k^{(2)}\right) + v_k.$$

The system parameters are $m = 0.5$kg, $l = 0.5$m, $\chi = 0.25$, $k_1 = -49.5$, $k_2 = -167.5$, sampling period $T = 0.01$s, and $\kappa = 0.5$N/m. The variances of w_k and v_k are 0.25 and 9×10^{-4}, respectively. Prob$\{\upsilon_k = 1\} = 0.8$. The saturation level is 0.2. The transmission threshold is set to be $\varsigma = 0.002$. The initial states are uniformly distributed over [0.5,1.5]. $\varepsilon_i(i = 1,2,3,5,7)$ are selected as 0.1, and $\varepsilon_i(i = 4,6,8)$ are determined as 1. For all $i = 1, 2, \ldots, 8$, $\tilde{\varepsilon}_i = \varepsilon_i$.

In the fault-free case, Fig. 7.1 and Fig. 7.2 show the systems states and their estimates. Fig. 7.3 illustrates the real filtering errors and the bound calculated from Theorem 7.2. It can be seen that acceptable estimation performance is achieved.

When the inverted pendulum is subject to unexpected torques, additive faults may occur. Consider a fault f_k in the following form in $x_k^{(1)}$:

$$f_k = \begin{cases} -0.9, & \text{if } k \geq 26, \\ 0, & \text{otherwise.} \end{cases}$$

Fig. 7.4 depicts the actual fault and its estimate obtained from Theorem 7.3. It can be observed that the proposed filter could estimate the additive fault well.

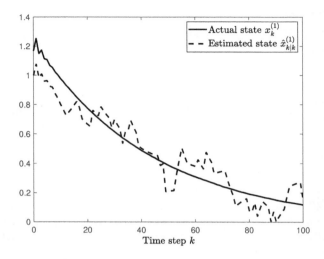

FIGURE 7.1: The state x_1 and its estimate

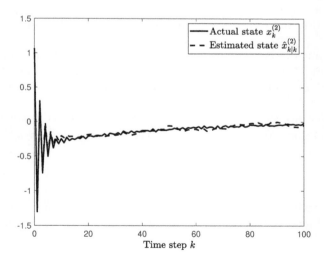

FIGURE 7.2: The state x_2 and its estimate

Illustrations 131

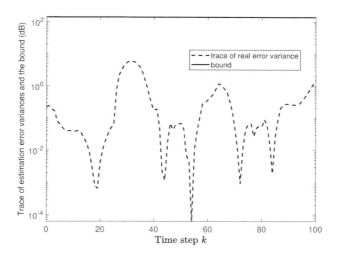

FIGURE 7.3: The estimation error and bound

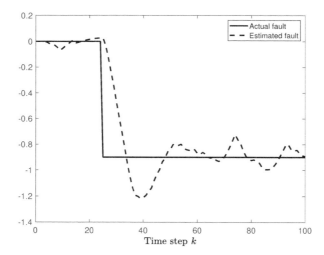

FIGURE 7.4: The fault and its estimate

7.6 Conclusions

In this chapter, the filtering problem has been investigated for a class of time-varying nonlinear systems with stochastic sensor saturations and event-triggered measurement transmissions. Special effort has been made to obtain an upper bound of the filtering error covariance and then minimize such an upper bound by solving two sets of discrete iterative matrix equations. The presented method has also been extended to estimate the additive faults. Future research topics would include the extension of our results to more complex systems such as nonlinear polynomial systems [6], delayed sensing systems [11], networked control systems and two-dimensional systems [87].

8

Finite-Horizon Quantized H_∞ Filter Design for Time-Varying Systems under Event-Triggered Transmissions

In the past decades, the filtering problem has been receiving persistent research attention from both control and signal processing communities. Among existing filtering approaches, the H_∞ filtering has been extensively studied to guarantee a bound of the worst-case estimation error where the information on the disturbances statistics is not required. To deal with the H_∞ filtering/control problems for *time-varying* systems over a finite horizon, differential/difference linear matrix inequality and recursive linear matrix inequality have been adopted in [49,143], and the backward recursive Riccati difference equation approach has also recently drawn specific interests [31,161].

The event-triggered transmission mechanism has recently gained much research attention owing to the need for reducing energy consumptions and extending the lifetimes of the services. Compared with the conventional clock-driven strategy, the outputs/inputs in an event-triggered scheme are released only when some conditions are violated in the event generator. As such, the event-triggered transmission strategy can reduce energy consumptions and extend the lifetimes of the services. The event-triggered minimum-variance filter has been studied in [172], where the PDFs of the states conditional on measurements have been approximated with Gaussian distributions. Modified Kalman filter for linear systems with event-triggered transmissions has been designed in [121,122,153], where the difference between the current measurement and the previously transmitted one has been assumed to be uniformly distributed. The event-triggered H_∞ filtering problem has also stirred some initial research attention in [67] for linear time-invariant systems.

It should be pointed out that, in almost all existing literature concerning event-triggered filtering problems, the time-varying and nonlinear behaviors, which are ubiquitous in practice, have not yet been taken into account despite their engineering significance. Moreover, in most reported results, the Euclidean norm (or the weighed Euclidean norm) of the measurement output has been utilized in the event generator. In practical, however, it is often the case that certain components of the measurement vector deserve more attention, which should be evaluated and transmitted separately/individually [115, 121, 122]. In such a case, the transmitted output signals are more

DOI: 10.1201/9781003309482-8

sensitive to the variations of measurements and this enables us to monitor the addressed systems more accurately with individual componentwise thresholds. Therefore, there is a practical need to address the H_∞ filtering problem for time-varying nonlinear systems with componentwise event-triggered transmissions.

Measurement outputs of practical systems are often subject to quantization effects due to limited bandwidth in a networked environment, and there has been a rich body of literature focusing on the filtering problem with quantization effects. Unfortunately, when it comes to the event-triggered scenario, the corresponding filtering problem with quantization effects has not gained adequate research attention yet especially when the system is time-varying and nonlinear. In [115], the event-triggered H_∞ filtering with output quantization has been investigated for linear time-invariant systems by using the LMI method. Therefore, it is of significance to consider the event-triggered H_∞ filter design for time-varying nonlinear systems with quantization effects over a finite horizon.

In this chapter, we aim to deal with the H_∞ filtering problem for a class of time-varying systems subject to event-triggered measurement transmissions and quantization effects. A componentwise event-triggered mechanism is put forward and a logarithmic quantizer is taken into account. Sufficient conditions are established to guarantee the prescribed H_∞ performance over a finite horizon. The filter gain is designed based on an H_2-type requirement, and the filter is capable of estimating the states over a finite horizon with a locally minimized cost. An illustrative example is provided to show the effectiveness of the proposed method in detecting possible faults in the system. The main novelty of the chapter lies in the following aspects: *1) a comprehensive model is considered which covers nonlinearities, componentwise event-triggered measurement transmissions, and quantization effects; 2) an H_∞ filtering performance is considered over a finite horizon in response to the time-varying nature of the addressed system; and 3) the finite-horizon H_∞ filter is designed via solving backward RDEs and the recursive algorithm is suitable for online computation.*

8.1 Problem Formulation

Consider the following class of discrete-time nonlinear systems:

$$\begin{cases} x_{k+1} = A_k x_k + g(x_k) + D_k w_k + E_k f_k, \\ y_k = C_k x_k + F_k v_k, \end{cases} \tag{8.1}$$

where $x_k \in \mathbb{R}^n$ is the state; $y_k \in \mathbb{R}^m$ is the measurement output; $f_k \in \mathbb{R}^t$ is the additive fault; $w_k \in \mathbb{R}^p$ and $v_k \in \mathbb{R}^q$ are the process noise and measurement noise, respectively. It is also assumed that $f_k, w_k, v_k \in l_2[0, N-1]$. A_k, C_k,

Problem Formulation

D_k, E_k, and F_k are known time-varying matrices with appropriate dimensions. The nonlinear function $g(\cdot)$ satisfies $g(0) = 0$ and the Lipschitz condition as follows:

$$\|g(x_1) - g(x_2)\|^2 \leq l_g \|x_1 - x_2\|^2, \tag{8.2}$$

where l_g is a known positive scalar.

Let $y_k^{(i)}$ ($i \in \{1, 2, \ldots, m\}$) be the ith component/element/entry of the vector y_k. The following componentwise transmission strategy is considered in this chapter: if the current measurement y_{k+j} and the previously transmitted one y_k satisfy the following condition:

$$\left| y_{k+j}^{(i)} - y_k^{(i)} \right| > \varsigma_i \left| y_{k+j}^{(i)} \right| \tag{8.3}$$

where ς_i is a given positive scalar, then the element $y_{k+j}^{(i)}$ would be transmitted. As a result, the event-triggered measurement $\hat{y}_k$ satisfies

$$\hat{y}_k^{(i)} = y_{k_{i,j}}^{(i)}, \quad k \in \{k_{i,j}, k_{i,j}+1, \cdots, k_{i,j+1}-1\} \tag{8.4}$$

where $k_{i,j}$ is the jth trigger instants of the ith entry.

For the quantization phenomenon, the logarithmic quantizer with time-invariant quantization levels is taken into account. For a vector $\hat{y}_k = \left[\hat{y}_k^{(1)}, \ldots, \hat{y}_k^{(m)} \right]^T$, the quantization process $q : \mathbb{R}^m \to \mathbb{R}^m$ is defined as:

$$q(\hat{y}_k) = \left[q_1(\hat{y}_k^{(1)}), \ldots, q_m(\hat{y}_k^{(m)}) \right]^T. \tag{8.5}$$

For all $s = 1, \ldots, m$, the quantization levels are given by:

$$\mathfrak{L}_s = \{\pm\alpha_{s,t}, \alpha_{s,t} = \beta_s^t \alpha_s, t = 0, \pm 1, \ldots\} \bigcup \{0\}, \tag{8.6}$$

where $\alpha_s > 0$, and $\beta_s \in (0, 1)$ is called the quantization density. The logarithmic quantizer is defined as

$$q_s(\hat{y}_k^{(s)}) = \begin{cases} \alpha_{s,t}, & \frac{1}{1+\delta_s}\alpha_{s,t} < \hat{y}_k^{(s)} < \frac{1}{1-\delta_s}\alpha_{s,t} \\ 0, & \hat{y}_k^{(s)} = 0 \\ -q_s(-\hat{y}_k^{(s)}), & \hat{y}_k^{(s)} < 0 \end{cases} \tag{8.7}$$

where $\delta_s = \dfrac{1-\beta_s}{1+\beta_s}$. It is straightforward to see that there exists $|\varrho_{k,s}| \leq \delta_s$ such that $q_s(\hat{y}_k^{(s)}) = (1 + \varrho_{k,s})\hat{y}_k^{(s)}$. Denoting $\Delta_k = \text{diag}\{\varrho_{k,1}, \ldots, \varrho_{k,m}\}$ and $\bar{\Delta} = \text{diag}\{\delta_1, \ldots, \delta_m\}$, then there exists a diagonal matrix $\tilde{\Delta}_k$ with $\|\tilde{\Delta}_k\| \leq 1$ such that $\Delta_k = \bar{\Delta}\tilde{\Delta}_k$. The quantized signal $\tilde{y}_k$ is written as

$$\tilde{y}_k = q(\hat{y}_k) = (I + \Delta_k)\hat{y}_k. \tag{8.8}$$

Remark 8.1 *The Lipschitz condition (8.2) applies to some closed-loop situations as well. If the nonlinear term in (8.1) is $g(x_k, u_k)$ where u_k is the amplitude-bounded control input with $\|u_k\| \leq \bar{u}$, then we could assume that the following inequality holds for all the $\|b\| \leq \bar{u}$:*

$$\|g(a_1, b) - g(a_2, b)\|^2 \leq l_g \|a_1 - a_2\|^2, \qquad (8.9)$$

which means that the addressed nonlinearity satisfies local Lipschitz condition with respect to the states.

Remark 8.2 *Event-triggered transmissions and quantization effects are considered in this chapter. A componentwise event-triggered strategy (8.3) is introduced inspired by the transmission mechanisms in [115, 121, 122]. In the proposed strategy, every entry of the output is evaluated and the individual thresholds ς_i can be adjusted based on the significance of the corresponding element. This triggering mechanism allows us to monitor the system dynamics more accurately and brings in more flexibility in the design of the event generator.*

For system (8.1), consider a filter of the following structure:

$$\hat{x}_{k+1} = A_k \hat{x}_k + g(\hat{x}_k) + K_k (\tilde{y}_k - C_k \hat{x}_k) \qquad (8.10)$$

where $\hat{x}_k \in \mathbb{R}^n$ is the estimate of x_k at time step k, and K_k is the filter gain to be determined.

Denoting the estimation error as $e_k = x_k - \hat{x}_k$, it can be obtained from (8.1), (8.8) and (8.10) that:

$$\begin{aligned} e_{k+1} =& A_k e_k + g(x_k) - g(\hat{x}_k) + D_k w_k + E_k f_k \\ &- K_k \left[(I + \Delta_k) \hat{y}_k - C_k \hat{x}_k\right]. \end{aligned} \qquad (8.11)$$

Letting $s_k = \hat{y}_k - y_k$ and adding a zero term $K_k(I + \Delta_k)y_k - K_k(I + \Delta_k)y_k$ to the right-hand side of (8.11), we have

$$\begin{aligned} e_{k+1} =& g(x_k) - g(\hat{x}_k) + (A_k - K_k C_k)e_k + D_k w_k + E_k f_k \\ &- K_k(I + \Delta_k) s_k - K_k \Delta_k C_k x_k - K_k(I + \Delta_k) F_k v_k. \end{aligned} \qquad (8.12)$$

Consider an augmented state $\xi_k = \begin{bmatrix} e_k^T, x_k^T \end{bmatrix}^T$ and define $\varpi_k = \begin{bmatrix} w_k^T, v_k^T, f_k^T \end{bmatrix}^T$. Then, it follows from (8.1) and (8.12) that:

$$\begin{aligned} \xi_{k+1} =& \tilde{g}(\xi_k) + \tilde{E}_k \varpi_k + \tilde{C}_k \xi_k + \tilde{F}_k s_k \\ &+ \tilde{F}_k \Delta_k F_k H_2 \varpi_k + \tilde{F}_k \Delta_k C_k H_1 \xi_k + \tilde{F}_k \Delta_k s_k, \end{aligned} \qquad (8.13)$$

where

$$\tilde{g}(\xi_k) = \begin{bmatrix} g(x_k) - g(\hat{x}_k) \\ g(x_k) \end{bmatrix}, \quad \tilde{C}_k = \begin{bmatrix} A_k - K_k C_k & 0 \\ 0 & A_k \end{bmatrix},$$

$$\tilde{E}_k = \begin{bmatrix} D_k & -K_k F_k & E_k \\ D_k & 0 & E_k \end{bmatrix}, \quad \tilde{F}_k = \begin{bmatrix} -K_k \\ 0 \end{bmatrix},$$

$$H_1 = [0, I], \quad H_2 = [0, I, 0].$$

Filter Design

The aim of this chapter is to design the sequence of filter parameter matrices K_k such that the estimation error satisfies the following finite-horizon requirement:

$$J = \sum_{k=0}^{N-1} \|e_k\|^2 - \gamma^2 \sum_{k=0}^{N-1} \|\varpi_k\|^2 - \gamma^2 \xi_0^T R \xi_0 < 0 \quad (\{\varpi_k\}, \xi_0 \neq 0) \tag{8.14}$$

where $R > 0$ is a weighting matrix and $\gamma > 0$ is the given attenuation level.

Remark 8.3 *The H_∞ filtering problem with quantization effects has been investigated for linear time-invariant systems over an infinite horizon in [115]. However, the LMI method adopted in [115] cannot be applied to time-varying nonlinear systems. For the addressed system (8.1), it makes more sense to focus on the H_∞ filtering problem over a finite horizon, which refers to the attenuation of the influences from external disturbances over a given interval $[0, N-1]$.*

8.2 Filter Design

Let us rewrite (8.13) as follows:

$$\xi_{k+1} = \tilde{C}_k \xi_k + \tilde{D}_k \hat{w}_k \tag{8.15}$$

where

$$\hat{w}_k = \left[\varpi_k^T, \varepsilon_{1k} \tilde{g}^T(\xi_k), \varepsilon_{2k}(\tilde{\Delta}_k F_k H_2 \varpi_k)^T, \right.$$

$$\left. \varepsilon_{3k}(\tilde{\Delta}_k C_k H_1 \xi_k)^T, \varepsilon_{4k}(\tilde{\Delta}_k s_k)^T, \varepsilon_{5k} s_k^T \right]^T,$$

$$\tilde{D}_k = \left[\tilde{E}_k, \varepsilon_{1k}^{-1} I, \varepsilon_{2k}^{-1} \tilde{F}_k \bar{\Delta}, \varepsilon_{3k}^{-1} \tilde{F}_k \bar{\Delta}, \varepsilon_{4k}^{-1} \tilde{F}_k \bar{\Delta}, \varepsilon_{5k}^{-1} \tilde{F}_k \right],$$

and ε_{ik} ($i = 1, \ldots, 5$) are known positive scalars.

For system (8.15), define the following performance requirement:

$$\bar{J} = \sum_{k=0}^{N-1} (\|e_k\|^2 - \gamma^2 \|\hat{w}_k\|^2) - \gamma^2 \xi_0^T R \xi_0 + \gamma^2 \sum_{k=0}^{N-1} (l_g \|\varepsilon_{1k}$$
$$\times \xi_k\|^2 + \|\varepsilon_{2k} F_k H_2 \varpi_k\|^2 + \|\varepsilon_{3k} C_k H_1 \xi_k\|^2 + \|\varepsilon_{4k} \Lambda$$
$$\times y_k\|^2 + \|\varepsilon_{5k} \Lambda y_k\|^2) < 0 \quad (\{\varpi_k\}, \xi_0 \neq 0), \tag{8.16}$$

where $\Lambda = \text{diag}\{\varsigma_1, \ldots, \varsigma_m\}$.

Theorem 8.1 *Considering requirements (8.14) and (8.16), we have*

$$J \leq \bar{J}.$$

Proof From the definitions of J and $\bar{J}$, it follows that

$$\bar{J} - J = \sum_{k=0}^{N-1} \gamma^2(\varepsilon_{4k}^2 + \varepsilon_{5k}^2)(\|\Lambda y_k\|^2 - \|s_k\|^2)$$

$$+ \sum_{k=0}^{N-1} \gamma^2(l_g\|\varepsilon_{1k}\xi_k\|^2 - \|\varepsilon_{1k}\tilde{g}(\xi_k)\|^2)$$

$$+ \sum_{k=0}^{N-1} \gamma^2(\|\varepsilon_{2k}F_kH_2\varpi_k\|^2 - \|\varepsilon_{2k}\tilde{\Delta}_kF_kH_2\varpi_k\|^2 + \|\varepsilon_{3k}C_kH_1\xi_k\|^2$$

$$- \|\varepsilon_{3k}\tilde{\Delta}_kC_kH_1\xi_k\|^2 + \|\varepsilon_{4k}s_k\|^2 - \|\varepsilon_{4k}\tilde{\Delta}_k s_k\|^2). \tag{8.17}$$

Based on the definition of $\tilde{\Delta}_k$, (8.17) can be written as

$$\bar{J} - J = \sum_{k=0}^{N-1} \gamma^2(\varepsilon_{4k}^2 + \varepsilon_{5k}^2)(\|\Lambda y_k\|^2 - \|s_k\|^2)$$

$$+ \sum_{k=0}^{N-1} \gamma^2(l_g\|\varepsilon_{1k}\xi_k\|^2 - \|\varepsilon_{1k}\tilde{g}(\xi_k)\|^2)$$

$$+ \sum_{k=0}^{N-1} \gamma^2(\|\varepsilon_{2k}(I - \tilde{\Delta}_k\tilde{\Delta}_k^T)^{1/2}F_kH_2\varpi_k\|^2$$

$$+ \|\varepsilon_{3k}(I - \tilde{\Delta}_k\tilde{\Delta}_k^T)^{1/2}C_kH_1\xi_k\|^2$$

$$+ \|\varepsilon_{4k}(I - \tilde{\Delta}_k\tilde{\Delta}_k^T)^{1/2}s_k\|^2). \tag{8.18}$$

With the event-triggered strategy (8.3) and the Lipschitz condition (8.2), we have that

$$\|\tilde{g}(\xi_k)\|^2 \le l_g\|\xi_k\|^2, \quad \|s_k\|^2 \le \|\Lambda y_k\|^2,$$

and then it follows directly from (8.18) that $J \le \bar{J}$.

With Theorem 8.1, it can be concluded that $J < 0$ can be ensured if $\bar{J} < 0$. A set of backward RDEs will be employed in the next theorem to guarantee that $\bar{J} < 0$.

Theorem 8.2 *Consider the system (8.1) with event-triggered transmissions (8.3) and quantization effects (8.5). Let the disturbance rejection attenuation level $\gamma > 0$ and the positive symmetric matrix R be given. The requirement (8.16) is achieved if there exist positive scalars ε_{ik} for $i = 1, 2, \ldots, 5$ and $k = 0, 1, \ldots, N-1$ such that the following discrete Riccati difference equation*

$$\begin{cases} P_k = \tilde{C}_k^T P_{k+1}\tilde{C}_k + H_3^T H_3 + \gamma^2 l_g \varepsilon_{1k}^2 I + \gamma^2 H_1^T C_k^T [\varepsilon_{3k}^2 I \\ \quad + (\varepsilon_{4k}^2 + \varepsilon_{5k}^2)\Lambda^T\Lambda]C_kH_1 + [\tilde{C}_k^T P_{k+1}\tilde{D}_k + \gamma^2(\varepsilon_{4k}^2 + \varepsilon_{5k}^2) \\ \quad \times H_1^T C_k^T\Lambda^T\Lambda F_kH_2H_4]\Omega_k^{-1}[\tilde{C}_k^T P_{k+1}\tilde{D}_k \\ \quad + \gamma^2(\varepsilon_{4k}^2 + \varepsilon_{5k}^2)H_1^T C_k^T\Lambda^T\Lambda F_kH_2H_4]^T, \\ P_N = 0, \end{cases} \tag{8.19}$$

Filter Design   139

has a solution satisfying

$$\begin{cases} \Omega_k = \gamma^2 I - \tilde{D}_k^T P_{k+1} \tilde{D}_k - \gamma^2 H_4^T H_2^T F_k^T \left[\varepsilon_{2k}^2 I + (\varepsilon_{4k}^2 + \varepsilon_{5k}^2) \Lambda^T \Lambda \right] \\ \quad \times F_k H_2 H_4 > 0, \\ P_0 < \gamma^2 R, \end{cases} \quad (8.20)$$

where $H_3 = [I, 0]$ *and* $H_4 = [I, 0, 0, 0, 0, 0]$.

Proof *Denote*

$$U_k = \xi_{k+1}^T P_{k+1} \xi_{k+1} - \xi_k^T P_k \xi_k. \quad (8.21)$$

Based on (8.15), we have

$$U_k = \xi_k^T (\tilde{C}_k^T P_{k+1} \tilde{C}_k - P_k) \xi_k + 2\xi_k^T \tilde{C}_k^T P_{k+1} \tilde{D}_k \hat{w}_k \\ + \hat{w}_k^T \tilde{D}_k^T P_{k+1} \tilde{D}_k \hat{w}_k. \quad (8.22)$$

Adding the zero term

$$\|e_k\|^2 - \gamma^2 (\|\hat{w}_k\|^2 - l_g \|\varepsilon_{1k} \xi_k\|^2 - \|\varepsilon_{2k} F_k H_2 \varpi_k\|^2 \\ - \|\varepsilon_{3k} C_k H_1 \xi_k\|^2 - \|\varepsilon_{4k} \Lambda y_k\|^2 - \|\varepsilon_{5k} \Lambda y_k\|^2) \\ - \Big[\|e_k\|^2 - \gamma^2 (\|\hat{w}_k\|^2 - l_g \|\varepsilon_{1k} \xi_k\|^2 - \|\varepsilon_{2k} F_k H_2 \varpi_k\|^2 \\ - \|\varepsilon_{3k} C_k H_1 \xi_k\|^2 - \|\varepsilon_{4k} \Lambda y_k\|^2 - \|\varepsilon_{5k} \Lambda y_k\|^2) \Big]$$

to the right hand side of (8.22), we have

$$U_k = \xi_k^T \Big\{ \tilde{C}_k^T P_{k+1} \tilde{C}_k + H_3^T H_3 + \gamma^2 l_g \varepsilon_{1k}^2 I + \gamma^2 H_1^T C_k^T \big[\varepsilon_{3k}^2 \\ \times I + (\varepsilon_{4k}^2 + \varepsilon_{5k}^2) \Lambda^T \Lambda \big] C_k H_1 - P_k \Big\} \xi_k + 2\xi_k^T \big[\tilde{C}_k^T \\ \times P_{k+1} \tilde{D}_k + \gamma^2 (\varepsilon_{4k}^2 + \varepsilon_{5k}^2) H_1^T C_k^T \Lambda^T \Lambda F_k H_2 H_4 \big] \hat{w}_k \\ + \hat{w}_k^T \Big\{ \tilde{D}_k^T P_{k+1} \tilde{D}_k + \gamma^2 H_4^T H_2^T F_k^T \big[\varepsilon_{2k}^2 I + (\varepsilon_{4k}^2 \\ + \varepsilon_{5k}^2) \Lambda^T \Lambda \big] F_k H_2 H_4 - \gamma^2 I \Big\} \hat{w}_k - \big[\|e_k\|^2 - \gamma^2 \\ \times (\|\hat{w}_k\|^2 - l_g \|\varepsilon_{1k} \xi_k\|^2 - \|\varepsilon_{2k} F_k H_2 \varpi_k\|^2 - \|\varepsilon_{3k} C_k \\ \times H_1 \xi_k\|^2 - \|\varepsilon_{4k} \Lambda y_k\|^2 - \|\varepsilon_{5k} \Lambda y_k\|^2) \big]. \quad (8.23)$$

Completing the squares with respect to $\hat{w}_k$ yields

$$\begin{aligned}U_k =& \xi_k^T \Big\{ \tilde{C}_k^T P_{k+1} \tilde{C}_k + H_3^T H_3 + \gamma^2 l_g \varepsilon_{1k}^2 I + \gamma^2 H_1^T C_k^T \big[\varepsilon_{3k}^2 \\ & \times I + (\varepsilon_{4k}^2 + \varepsilon_{5k}^2) \Lambda^T \Lambda \big] C_k H_1 + [\tilde{C}_k^T P_{k+1} \tilde{D}_k + \gamma^2 (\varepsilon_{4k}^2 \\ & + \varepsilon_{5k}^2) H_1^T C_k^T \Lambda^T \Lambda F_k H_2 H_4] \Omega_k^{-1} [\tilde{C}_k^T P_{k+1} \tilde{D}_k + \gamma^2 (\varepsilon_{4k}^2 \\ & + \varepsilon_{5k}^2) H_1^T C_k^T \Lambda^T \Lambda F_k H_2 H_4]^T - P_k \Big\} \xi_k - (\hat{w}_k - \hat{w}_k^*)^T \\ & \times \Omega_k (\hat{w}_k - \hat{w}_k^*) - \big[\|e_k\|^2 - \gamma^2 (\|\hat{w}_k\|^2 - l_g \|\varepsilon_{1k} \xi_k\|^2 \\ & - \|\varepsilon_{2k} F_k H_2 \varpi_k\|^2 - \|\varepsilon_{3k} C_k H_1 \xi_k\|^2 - \|\varepsilon_{4k} \Lambda y_k\|^2 \\ & - \|\varepsilon_{5k} \Lambda y_k\|^2) \big], \end{aligned} \qquad (8.24)$$

where

$$\hat{w}_k^* = \Omega_k^{-1} [\tilde{C}_k^T P_{k+1} \tilde{D}_k + \gamma^2 (\varepsilon_{4k}^2 + \varepsilon_{5k}^2) H_1^T C_k^T \Lambda^T \Lambda F_k H_2 H_4]^T \xi_k. \qquad (8.25)$$

Taking the sum on both sides of (8.24) from 0 to $N-1$ and considering (8.19), we have

$$\begin{aligned} & \xi_N^T P_N \xi_N - \xi_0^T P_0 \xi_0 \\ =& - \sum_{k=0}^{N-1} (\hat{w}_k - \hat{w}_k^*)^T \Omega_k (\hat{w}_k - \hat{w}_k^*) - \sum_{k=0}^{N-1} \Big[\|e_k\|^2 - \gamma^2 \\ & \times \big(\|\hat{w}_k\|^2 - l_g \|\varepsilon_{1k} \xi_k\|^2 - \|\varepsilon_{2k} F_k H_2 \varpi_k\|^2 - \|\varepsilon_{3k} C_k H_1 \\ & \times \xi_k\|^2 - \|\varepsilon_{4k} \Lambda y_k\|^2 - \|\varepsilon_{5k} \Lambda y_k\|^2 \big) \Big]. \end{aligned} \qquad (8.26)$$

From the facts that $\Omega_k > 0$, $P_N = 0$, $P_0 < \gamma^2 R$ and $\xi_0 \neq 0$, we have

$$\begin{aligned} \bar{J} =& \sum_{k=0}^{N-1} (\|e_k\|^2 - \gamma^2 \|\hat{w}_k\|^2) - \gamma^2 \xi_0^T R \xi_0 + \gamma^2 \sum_{k=0}^{N-1} (l_g \|\varepsilon_{1k} \\ & \times \xi_k\|^2 + \|\varepsilon_{2k} F_k H_2 \varpi_k\|^2 + \|\varepsilon_{3k} C_k H_1 \xi_k\|^2 \\ & + \|\varepsilon_{4k} \Lambda y_k\|^2 + \|\varepsilon_{5k} \Lambda y_k\|^2) \\ =& -\xi_0^T (\gamma^2 R - P_0) \xi_0 - \sum_{k=0}^{N-1} (\hat{w}_k - \hat{w}_k^*)^T \Omega_k (\hat{w}_k - \hat{w}_k^*) < 0, \end{aligned} \qquad (8.27)$$

which concludes the proof.

To design the desired filter gain K_k, we now consider the worst-case disturbance $\hat{w}_k = \hat{w}_k^*$ given in (8.25). The system (8.15) can be rearranged as

$$\begin{aligned} \xi_{k+1} =& \Big\{ \bar{A}_k + \tilde{D}_k \Omega_k^{-1} [\tilde{C}_k^T P_{k+1} \tilde{D}_k + \gamma^2 (\varepsilon_{4k}^2 + \varepsilon_{5k}^2) H_1^T C_k^T \\ & \times \Lambda^T \Lambda F_k H_2 H_4]^T \Big\} \xi_k - z_k, \end{aligned} \qquad (8.28)$$

Filter Design 141

where

$$\bar{A}_k = \begin{bmatrix} A_k & 0 \\ 0 & A_k \end{bmatrix}, \quad z_k = \begin{bmatrix} K_k C_k e_k \\ 0 \end{bmatrix}.$$

Associated with the design of a locally optimal filter, a cost function is defined as follows:

$$\Phi = \sum_{k=0}^{N-1} (\|e_k\|^2 + \|z_k\|^2). \tag{8.29}$$

The adoption of the function (8.29) is along the similar line of determining the H_2-type criteria in [31, 161]. This function is related to the sum of the inputs and outputs of the filter. In fact, when the measurements are free of additive noises, event-triggered transmissions and quantizations, the term $\|z_k\|^2 = \|K_k(\tilde{y}_k - C_k \hat{x}_k)\|^2$ can be seen as the input energy of the filter at time step k. Subsequently, the filter gain will be designed to minimize the function (8.29) under the constraint of the H_∞ requirement (8.14).

Theorem 8.3 *Consider the system (8.1) with event-triggered transmissions (8.3) and quantization effects (8.5). Let the disturbance rejection attenuation level $\gamma > 0$ and the positive symmetric matrix R be given. For the worst-case disturbance $\{\hat{w}_k^*\}_{k=0,1,\ldots,N-1}$, the requirement (8.16) is satisfied if there exist positive scalars ε_{ik} ($i = 1, 2, \ldots, 5; k = 0, 1, \ldots, N-1$) such that discrete Riccati difference equation (8.19) has a solution (P_k, K_k) satisfying (8.20) and the following equation*

$$\begin{cases} Q_k = \left\{ \bar{A}_k + \tilde{D}_k \Omega_k^{-1} [\tilde{C}_k^T P_{k+1} \tilde{D}_k + \gamma^2 (\varepsilon_{4k}^2 + \varepsilon_{5k}^2) H_1^T C_k^T \right. \\ \qquad \left. \times \Lambda^T \Lambda F_k H_2 H_4 \right]^T \bigg\}^T Q_{k+1} \left\{ \bar{A}_k + \tilde{D}_k \Omega_k^{-1} [\tilde{C}_k^T P_{k+1} \right. \\ \qquad \left. \times \tilde{D}_k + \gamma^2 (\varepsilon_{4k}^2 + \varepsilon_{5k}^2) H_1^T C_k^T \Lambda^T \Lambda F_k H_2 H_4 \right]^T \bigg\} \\ \qquad + H_3^T H_3 - \bar{A}_k^T Q_{k+1} \Sigma_k^{-1} Q_{k+1} \bar{A}_k - \bar{A}_k^T Q_{k+1} \Sigma_k^{-1} \\ \qquad \times Q_{k+1} \tilde{D}_k \Omega_k^{-1} [\tilde{C}_k^T P_{k+1} \tilde{D}_k + \gamma^2 (\varepsilon_{4k}^2 + \varepsilon_{5k}^2) H_1^T C_k^T \\ \qquad \times \Lambda^T \Lambda F_k H_2 H_4]^T - [\tilde{C}_k^T P_{k+1} \tilde{D}_k + \gamma^2 (\varepsilon_{4k}^2 + \varepsilon_{5k}^2) \\ \qquad \times H_1^T C_k^T \Lambda^T \Lambda F_k H_2 H_4] \Omega_k^{-1} \tilde{D}_k^T Q_{k+1} \Sigma_k^{-1} Q_{k+1} \bar{A}_k, \\ Q_N = 0, \end{cases} \tag{8.30}$$

has a solution (Q_k, K_k) satisfying

$$\Sigma_k = Q_{k+1} + I > 0. \tag{8.31}$$

In this case, the filter gain is given by

$$K_k = H_3 \Sigma_k^{-1} Q_{k+1} \bar{A}_k (C_k H_3)^\dagger. \tag{8.32}$$

Proof Based on Theorem 8.2, it can be concluded that the H_∞ requirement can be satisfied if the discrete Riccati difference equation (8.19) has a solution (P_k, K_k) satisfying (8.20) for every $k = 0, 1, \ldots, N-1$, and then the filter gain is to be designed with the worst-case disturbance $\hat{w}_k^*$ given in (8.25). For this purpose, denote

$$V_k = \xi_{k+1}^T Q_{k+1} \xi_{k+1} - \xi_k^T Q_k \xi_k. \tag{8.33}$$

Since system (8.15) with the worst-case disturbance $\hat{w}_k^*$ can be written as (8.28), we have

$$\begin{aligned}V_k =& \xi_k^T \Big\{ \Big\{ \bar{A}_k + \tilde{D}_k \Omega_k^{-1} [\tilde{C}_k^T P_{k+1} \tilde{D}_k + \gamma^2 (\varepsilon_{4k}^2 + \varepsilon_{5k}^2) H_1^T C_k^T \\ & \times \Lambda^T \Lambda F_k H_2 H_4]^T \Big\}^T Q_{k+1} \Big\{ \bar{A}_k + \tilde{D}_k \Omega_k^{-1} [\tilde{C}_k^T P_{k+1} \tilde{D}_k \\ & + \gamma^2 (\varepsilon_{4k}^2 + \varepsilon_{5k}^2) H_1^T C_k^T \Lambda^T \Lambda F_k H_2 H_4]^T \Big\} - Q_k \Big\} \xi_k \\ & - 2\xi_k^T \Big\{ \bar{A}_k + \tilde{D}_k \Omega_k^{-1} [\tilde{C}_k^T P_{k+1} \tilde{D}_k + \gamma^2 (\varepsilon_{4k}^2 + \varepsilon_{5k}^2) H_1^T \\ & \times C_k^T \Lambda^T \Lambda F_k H_2 H_4]^T \Big\}^T Q_{k+1} z_k + z_k^T Q_{k+1} z_k, \end{aligned} \tag{8.34}$$

and therefore

$$\begin{aligned}V_k =& \xi_k^T \Big\{ \Big\{ \bar{A}_k + \tilde{D}_k \Omega_k^{-1} [\tilde{C}_k^T P_{k+1} \tilde{D}_k + \gamma^2 (\varepsilon_{4k}^2 + \varepsilon_{5k}^2) H_1^T C_k^T \\ & \times \Lambda^T \Lambda F_k H_2 H_4]^T \Big\}^T Q_{k+1} \Big\{ \bar{A}_k + \tilde{D}_k \Omega_k^{-1} [\tilde{C}_k^T P_{k+1} \tilde{D}_k \\ & + \gamma^2 (\varepsilon_{4k}^2 + \varepsilon_{5k}^2) H_1^T C_k^T \Lambda^T \Lambda F_k H_2 H_4]^T \Big\} - Q_k \Big\} \xi_k \\ & - 2\xi_k^T \Big\{ \bar{A}_k + \tilde{D}_k \Omega_k^{-1} [\tilde{C}_k^T P_{k+1} \tilde{D}_k + \gamma^2 (\varepsilon_{4k}^2 + \varepsilon_{5k}^2) H_1^T \\ & \times C_k^T \Lambda^T \Lambda F_k H_2 H_4]^T \Big\}^T Q_{k+1} z_k + z_k^T Q_{k+1} z_k + \|e_k\|^2 \\ & + \|z_k\|^2 - \|e_k\|^2 - \|z_k\|^2. \end{aligned} \tag{8.35}$$

Noticing that $e_k = H_3 \xi_k$ and $\Sigma_k = Q_{k+1} + I$, (8.35) can be written as

$$\begin{aligned}V_k =& \xi_k^T \Big\{ \Big\{ \bar{A}_k + \tilde{D}_k \Omega_k^{-1} [\tilde{C}_k^T P_{k+1} \tilde{D}_k + \gamma^2 (\varepsilon_{4k}^2 + \varepsilon_{5k}^2) H_1^T C_k^T \Lambda^T \Lambda F_k H_2 H_4]^T \Big\}^T \\ & \times Q_{k+1} \Big\{ \bar{A}_k + \tilde{D}_k \Omega_k^{-1} [\tilde{C}_k^T P_{k+1} \tilde{D}_k + \gamma^2 (\varepsilon_{4k}^2 + \varepsilon_{5k}^2) H_1^T C_k^T \Lambda^T \Lambda F_k H_2 \\ & \times H_4]^T \Big\} + H_3^T H_3 - Q_k \Big\} \xi_k - 2\xi_k^T \Big\{ \bar{A}_k + \tilde{D}_k \Omega_k^{-1} [\tilde{C}_k^T P_{k+1} \tilde{D}_k + \gamma^2 (\varepsilon_{4k}^2 \\ & + \varepsilon_{5k}^2) H_1^T C_k^T \Lambda^T \Lambda F_k H_2 H_4]^T \Big\}^T Q_{k+1} z_k + z_k^T \Sigma_k z_k - \|e_k\|^2 - \|z_k\|^2. \end{aligned} \tag{8.36}$$

Filter Design

By completing the squares with respect to z_k, we obtain

$$\begin{aligned}
V_k =& \xi_k^T \Big\{ \big\{ \bar{A}_k + \tilde{D}_k \Omega_k^{-1} [\tilde{C}_k^T P_{k+1} \tilde{D}_k + \gamma^2(\varepsilon_{4k}^2 + \varepsilon_{5k}^2) H_1^T C_k^T \\
& \times \Lambda^T \Lambda F_k H_2 H_4 \big]^T \big\}^T Q_{k+1} \big\{ \bar{A}_k + \tilde{D}_k \Omega_k^{-1} [\tilde{C}_k^T P_{k+1} \tilde{D}_k \\
& + \gamma^2(\varepsilon_{4k}^2 + \varepsilon_{5k}^2) H_1^T C_k^T \Lambda^T \Lambda F_k H_2 H_4 \big]^T \big\} + H_3^T H_3 \\
& - \bar{A}_k^T Q_{k+1} \Sigma_k^{-1} Q_{k+1} \bar{A}_k - \bar{A}_k^T Q_{k+1} \Sigma_k^{-1} Q_{k+1} \tilde{D}_k \Omega_k^{-1} \\
& \times [\tilde{C}_k^T P_{k+1} \tilde{D}_k + \gamma^2(\varepsilon_{4k}^2 + \varepsilon_{5k}^2) H_1^T C_k^T \Lambda^T \Lambda F_k H_2 H_4]^T \\
& - [\tilde{C}_k^T P_{k+1} \tilde{D}_k + \gamma^2(\varepsilon_{4k}^2 + \varepsilon_{5k}^2) H_1^T C_k^T \Lambda^T \Lambda F_k H_2 H_4] \\
& \times \Omega_k^{-1} \tilde{D}_k^T Q_{k+1} \Sigma_k^{-1} Q_{k+1} \bar{A}_k - Q_k \Big\} \xi_k + (z_k - z_k^*)^T \\
& \times \Sigma_k (z_k - z_k^*) - \|e_k\|^2 - \|z_k\|^2,
\end{aligned} \quad (8.37)$$

where

$$z_k^* = \Sigma_k^{-1} Q_{k+1} \bar{A}_k \xi_k. \quad (8.38)$$

Then, it follows directly from (8.30) that

$$\Phi = \xi_0^T Q_0 \xi_0 + \sum_{k=0}^{N-1} (z_k - z_k^*)^T \Sigma_k (z_k - z_k^*). \quad (8.39)$$

It is obvious that Φ can be suppressed if $z_k = z_k^*$, i.e.,

$$K_k C_k H_3 = H_3 \Sigma_k^{-1} Q_{k+1} \bar{A}_k. \quad (8.40)$$

By resorting to the Moore-Penrose pseudo inverse, the filter gain can be determined as (8.32), and the proof is now complete.

The proposed algorithm can be concluded as follows:

Step 1. Set $k = T$, then $P_{T+1} = Q_{T+1} = 0$ are available.

Step 2. Determine positive scalars ε_{ik}s and calculate the matrices Ω_k, Σ_k by (8.20) and (8.31), respectively.

Step 3. If Ω_k and Σ_k are both positive definite, then we can obtain the filter gain K_k by (8.32) and step to the next procedure, else jump to Step 6.

Step 4. Solve the backward RDEs of (8.19) and (8.30) to get P_k and Q_k, respectively.

Step 5. If $k \neq 0$, set $k = k - 1$ and go back to Step 2, else turn to the next step.

Step 6. If $\Omega_k < 0$, or $\Sigma_k < 0$, or $P_0 \geq \gamma^2 R$, this algorithm is infeasible. Stop.

For the purpose of illustrating the proposed filter design algorithm in the fault detection problem, we utilize the following signal as the residual:

$$r_k = \tilde{y}_k - C_k \hat{x}_k. \qquad (8.41)$$

It can be seen that the possible fault will increase the Euclidean norm of the signal, so r_k can reflect the abnormal changes in the system efficiently. In the next section, a simulation example will be used to illustrate the effectiveness of the proposed filter.

Remark 8.4 *In the two sets of RDEs, the matrix Λ quantities the effects from the proposed componentwise event-triggered strategy, and the matrix $\bar{\Delta}$ reflects the influences of the quantization. The scalar l_g represents the nonlinearities, and the matrix $\tilde{D}_k$ stands for the influences of the faults and noises. The parameters ε_{ik}s can be determined to balance the impacts of the external disturbance and the influences of the nonlinearity, quantization, and event-triggered transmission in the H_∞ requirement, and how to properly select ε_{ik}s in a quantitative way is one of our future research directions. Sufficient conditions are obtained that can guarantee the predefined finite-horizon H_∞ requirement. The filter gain can be calculated by resorting to the pseudo inverse operator and solving the backward RDEs which locally minimizes the cost function (8.29). Since the RDEs can be solved recursively, the method is suitable for online applications. The second inequality in (8.20) is put forward to constrain the effects of the initial states on the filtering error and it can ensure that the solution of the established RDE is bounded. The computational complexity of the proposed algorithm is $O(N(4m + 2n + p + q + t)^3)$, and it is obvious that the computational burden of the proposed algorithm is polynomial with respect to the dimensions of the state and measurement, rather than exponential. As such, the proposed method can be applied to large scale systems. The developed method can be easily extended to the case where the event-triggered thresholds and quantization levels are time-varying.*

8.3 An Illustrative Example

Consider the following time-varying nonlinear system:

$$\begin{cases} x_{k+1}^{(1)} = 0.5 x_k^{(1)} \sin k - 0.7 x_k^{(2)} - 1.5 u_k + 0.01 w_k^{(1)} + f_k, \\ x_{k+1}^{(2)} = 0.15 \sin x_k^{(1)} + 0.4 x_k^{(2)} + 0.75 x_k^{(3)} \cos k + 0.01 w_k^{(2)}, \\ x_{k+1}^{(3)} = 0.2 x_k^{(1)} + 0.7 u_k x_k^{(2)} + 0.4 x_k^{(3)} + 0.01 w_k^{(3)}, \\ y_k^{(1)} = x_k^{(1)} + 0.01 v_k^{(1)}, \\ y_k^{(2)} = x_k^{(2)} + 0.01 v_k^{(2)}, \\ u_k = 0.2 \tilde{y}_k^{(1)} + 0.4 \tilde{y}_k^{(2)}, \end{cases}$$

An Illustrative Example

where $x_k = [x_k^{(1)}, x_k^{(2)}, x_k^{(3)}]^T \in \mathbb{R}^3$ is the system state, $y_k = [y_k^{(1)}, y_k^{(2)}]^T \in \mathbb{R}^2$ is the measurement output, $\tilde{y}_k = [\tilde{y}_k^{(1)}, \tilde{y}_k^{(2)}]^T \in \mathbb{R}^2$ is the transmitted output, $u_k \in \mathbb{R}$ is the control input, and $f_k \in \mathbb{R}$ is the additive fault.

The disturbances $w_k^{(i)}(i = 1, 2, 3)$ and $v_k^{(i)}(i = 1, 2)$ are mutually independent Gaussian distributed sequences with unity variances, and $\bar{u} = 0.1$. Every entry of the initial state is uniformly distributed over $[-0.03, 0.03]$. The event-triggered transmission thresholds are chosen as $\varsigma_1 = \varsigma_2 = 0.5$. Consider the parameters of the logarithmic quantizer as $\alpha_1 = \alpha_2 = 0.5$ and $\beta_1 = \beta_2 = 1/3$. The H_∞ performance level γ and the positive definite matrix R are given by $\gamma = 0.4$ and $R = 100I$, respectively. The scalars $\varepsilon_{ik}(i = 1, 2, \ldots, 5,$ and $k = 0, 1, \ldots, N-1)$ are determined as 2, 0.5, 1, 0.5, and 0.5 for all the time step k, respectively. N is set to be 40. It can be guaranteed that the sufficient conditions in Theorems 2-3 hold with such a group of parameters. When the system is fault-free and the initial state is $[5, 5, 5]^T$, the average Euclidean norm of the residuals obtained with the proposed filter is 1.4146×10^{-4}. When the effects of event-triggered transmissions and quantizations are neglected in the RDEs, the average Euclidean norm of the residuals is 2.2439×10^{-4}. So the proposed filter can better estimate the system states with significant quantization errors due to our specific efforts to handle the network-induced phenomena.

The additive fault f_k is in the following form:

$$f_k = \begin{cases} 0.1, & \text{if} \quad k \geq 25, \\ 0, & \text{otherwise}. \end{cases}$$

Figs. 8.1–8.3 plot the states and their estimates obtained from Theorem 8.3 under zero-initial condition. Fig. 8.4 illustrates the Euclidean norm of the residual. It can be seen that the proposed filter estimates the states well when there is no fault in the system (when $k < 25$), and the increase of the norm of the residual in the faulty case (when $k \geq 25$) could help us to detect the fault. The average Euclidean norm of the residuals obtained with the proposed filter after the fault occurs is 3.309×10^{-2}, which is much larger than that in the fault-free case. So the proposed filter is sensitive to the fault and can efficiently solve the fault detection problem. The actual attenuation level in the example is 0.3782, which satisfies the predefined finite-horizon requirement.

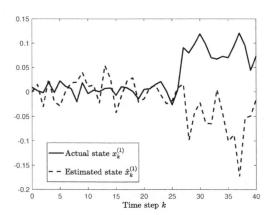

FIGURE 8.1: The state x_1 and its estimate

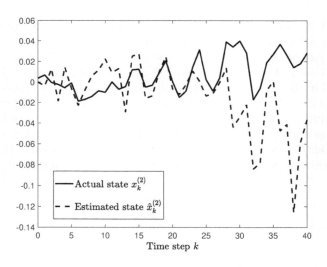

FIGURE 8.2: The state x_2 and its estimate

An Illustrative Example

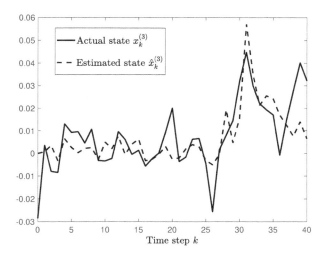

FIGURE 8.3: The state x_3 and its estimate

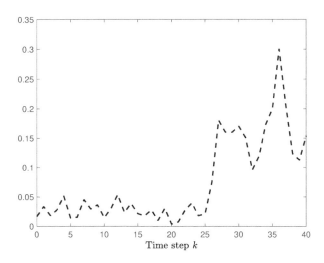

FIGURE 8.4: The Euclidean norm of the residual

8.4 Conclusion

The filtering problem has been considered for a class of time-varying systems with quantization effects and event-triggered measurement transmissions. A componentwise event-triggered transmission scheme has been put forward to reduce unnecessary signal transmissions and improve the energy efficiency. The output measurements have been assumed to be quantized by a logarithmic quantizer. An H_∞ requirement over a given finite horizon has been investigated. Two sets of Riccati-like matrix equations have been employed to ensure the predetermined H_∞ performance and design the filter, and the recursive algorithm is suitable for online computation. An illustrative example concerning fault detection has been presented to show the effectiveness of the proposed algorithm. Future research topics would include the extension of our results to more practical and complex systems such as fuzzy systems, uncertain systems, Markovian jump systems, etc.

9

Observer-Based Fault Diagnosis Schemes under Closed-Loop Control

In the past decades, much research attention has been paid to the fault detection and diagnosis (FDD) problems due to the increasing complexity and safety demands in modern industrial systems [62]. The FDD problems have been extensively studied in many practical systems such as photovoltaic panels, diesel engines, induction motors, underground cables, steam turbines, high-speed railways, etc. The FDD methods enable us to detect, isolate, and estimate possible faults such that severe performance deteriorations and disasters can be prevented as early as possible.

Among the existing FDD methods, observer-based methods have been studied in many scenarios. In particular, the classical Luenberger observer has been widely utilized in linear systems under certain completely observable condition owing to its simple structure and stable performance since it was proposed in 1971 [109]. Some modified Luenberger observers have been developed so as to cope with nonlinearities as well. A kind of robust observer has been established in [47] to detect and isolate faults with special structures of some significant parameters. When the addressed systems are subject to unknown disturbances, unknown input observer (UIO) methods have been presented to guarantee that the residuals are completely decoupled from disturbances [23]. The UIO techniques have been applied to various systems including uncertain systems, distributed systems, and singular systems.

Many real-world systems are under closed-loop control under the stability and robustness requirements. With properly designed control laws, the closed-loop controllers can adjust the control inputs on-line according to the system outputs. It should be pointed out that, in most reported FDD literature using observer-based methods, it has been implicitly assumed that the control input is exactly known at every time step. However, the closed-loop control input in practical systems may be unavailable to the FDD unit owing to many reasons such as large scales and distributed structures of the systems. In these situations, only the prescribed control law, rather than the actual control input, can be used in the FDD units. As a result, it would be interesting to examine how the feedback control would influence the performances of the FDD approaches, and this gives rise to the fundamental issue of evaluating the performance of existing schemes within a closed-loop framework. Furthermore, in the case that the FDD performance is indeed degraded by the feedback

DOI: 10.1201/9781003309482-9

control, it is of more engineering significance to modify the existing schemes to cater for the closed-loop control. So far, despite its clear engineering insight, such a closed-loop FDD performance evaluation problem has been largely overlooked due probably to the difficulties in mathematical formulation. It is, therefore, the main aim of this chapter to shorten such a gap by initializing the study on the FDD problem under closed-loop control.

In this chapter, the FDD problems are investigated for a class of closed-loop systems under PI control that is fairly popular in engineering practice. Some frequently used observer-based FDD approaches are examined within the closed-loop framework. To be more specific, the UIO-based approach is studied for a class of systems with unknown disturbances, and then both the Luenberger-observer-based method and the robust-observer-based FDD method are discussed in the disturbance-free case. It is shown that some methods are no longer applicable in the closed-loop case when only the control laws are available. Subsequently, the observers are redesigned such that the dynamics of the closed-loop residuals can be the same as those of the residuals achieved with control inputs. A simulation example is presented to show the effectiveness of the proposed method in the closed-loop situation. The main novelty of the chapter is twofold: 1) it is illustrated that the performance of the closed-loop residuals may be deteriorated by the effects of the feedback control; and 2) the structures of the observers can be adjusted and the parameters can be re-calculated in the closed-loop case such that the modified FDD methods are applicable in the closed-loop cases.

9.1 Unknown-Input-Observer Method

Consider the following class of linear discrete-time systems with unknown disturbances:

$$\begin{cases} x_{k+1} = Ax_k + Bu_k + Ed_k + f_k + w_k, \\ y_k = Cx_k + v_k, \end{cases} \quad (9.1)$$

where $x_k \in \mathbb{R}^n$ is the state; $u_k \in \mathbb{R}^p$ is the control input; $y_k \in \mathbb{R}^m$ is measurement output; $d_k \in \mathbb{R}^q$ is the unknown external disturbance; $f_k \in \mathbb{R}^n$ is the additive fault; $w_k \in \mathbb{R}^n$ and $v_k \in \mathbb{R}^m$ are the process noise and measurement noise, respectively; and A, B, C, and E are known matrices with appropriate dimensions.

For system (9.1), the UIO of the following structure is adopted as a discrete version of that in [23] when u_k is attainable at each time step:

$$\begin{cases} z_{k+1} = Fz_k + TBu_k + Ky_k, \\ \hat{x}_{k+1} = z_{k+1} + Hy_{k+1}, \end{cases} \quad (9.2)$$

where $z_k \in \mathbb{R}^n$ is the observer state and $\hat{x}_k \in \mathbb{R}^n$ is the state estimate. The parameters F, T, K, and H are provided as follows:

$$K = K_1 + K_2, \tag{9.3}$$
$$E = HCE, \tag{9.4}$$
$$T = I - HC, \tag{9.5}$$
$$F = A - HCA - K_1C, \tag{9.6}$$
$$K_2 = FH, \tag{9.7}$$

and the system matrix F is Schur stable (i.e., all the eigenvalues of F lie within the open unit disk). With the given UIO, the dynamics of estimation error $e_k = x_k - \hat{x}_k$ is governed by:

$$e_{k+1} = Fe_k - Hv_{k+1} - K_1 v_k + (I - HC)w_k + (I - HC)f_k. \tag{9.8}$$

The residual signal is taken as

$$r_k = y_k - C\hat{x}_k = Ce_k + v_k, \tag{9.9}$$

which is fully decoupled from the disturbance d_k. Necessary and sufficient conditions for (9.2) to be a UIO for the system defined by (9.1) are

1. rank(CE)=rank(E).
2. The pair $(C, A - HCA)$ is detectable.

Consider the discrete PI controller

$$u_k = Q_p y_k + Q_i \sum_{j=1}^{l} y_{k-j}, \tag{9.10}$$

where l, Q_p, and Q_i are predefined parameters that can ensure the stability of the closed-loop system. When only the control law is available to the observer (rather than the control input), substituting the control law into the original system (9.1) yields

$$\begin{cases} x_{k+1} = (A + BQ_pC)x_k + BQ_iC\sum_{j=1}^{l} x_{k-j} + Ed_k \\ \qquad\quad + f_k + w_k + BQ_p v_k + BQ_i \sum_{j=1}^{l} v_{k-j}, \\ y_k = Cx_k + v_k. \end{cases} \tag{9.11}$$

To apply the UIO method, define the following augmented state and measurement noise:

$$\bar{x}_k = [x_k^T, x_{k-1}^T, \ldots, x_{k-l}^T]^T, \quad \bar{v}_k = [v_k^T, v_{k-1}^T, \ldots, v_{k-l}^T]^T.$$

Then, the closed-loop system (9.11) can be rewritten in the following form:

$$\begin{cases} \bar{x}_{k+1} = \bar{A}\bar{x}_k + \bar{E}d_k + \bar{D}f_k + \bar{D}w_k + \bar{F}_1 \bar{v}_k, \\ y_k = \bar{C}\bar{x}_k + \bar{F}_2 \bar{v}_k, \end{cases} \tag{9.12}$$

where
$$\bar{A} = \begin{bmatrix} A+BQ_pC & BQ_iC & \cdots & BQ_iC \\ I & & & \\ & \ddots & & \\ & & I & 0 \end{bmatrix}, \quad \bar{F}_1 = \begin{bmatrix} BQ_p & BQ_i & \cdots & BQ_i \\ 0 & \cdots & \cdots & 0 \\ \vdots & \cdots & \cdots & \vdots \\ 0 & \cdots & \cdots & 0 \end{bmatrix},$$
$\bar{C} = [C, 0, \cdots, 0]$, $\bar{E} = [E^T, 0, \cdots, 0]^T$, $\bar{D} = [I, 0, \cdots, 0]^T$, $\bar{F}_2 = [I, 0, \cdots, 0]$.

For system (9.12), a closed-loop UIO of the following structure is adopted:
$$\begin{cases} \bar{z}_{k+1} = \bar{F}\bar{z}_k + \bar{K}y_k, \\ \hat{\bar{x}}_{k+1} = \bar{z}_{k+1} + \bar{H}y_{k+1}, \end{cases} \quad (9.13)$$

where $\bar{z}_k \in \mathbb{R}^{nl}$ is the observer state and $\hat{\bar{x}}_k \in \mathbb{R}^{nl}$ is the state estimate.

To guarantee that the closed-loop residual is still decoupled from the disturbance, we need to choose $\bar{H}$ such that
$$\bar{E} = \bar{H}\bar{C}\bar{E}. \quad (9.14)$$

Considering the structure of $\bar{E}$, the matrix $\bar{H}$ should be determined as
$$\bar{H} = [H^T, 0, \cdots, 0]^T \quad (9.15)$$

and, with $\bar{H}$ given in (9.15), we have
$$\bar{F} = \bar{A} - \bar{H}\bar{C}\bar{A} - \bar{K}_1\bar{C}$$
$$= \begin{bmatrix} (I-HC)(A+BQ_pC) & (I-HC)BQ_iC & \cdots & (I-HC)BQ_iC \\ I & & & \\ & \ddots & & \\ & & I & 0 \end{bmatrix} - \bar{K}_1\bar{C}.$$

To make the dynamics of the closed-loop residuals as similar as possible with that of residuals obtained with control inputs u_k, one needs to choose
$$\bar{K}_1 = [(K_1 + BQ_p - HCBQ_p)^T, 0, \cdots, 0]^T \quad (9.16)$$

and it then follows that
$$\bar{F} = \begin{bmatrix} F & (I-HC)BQ_iC & \cdots & (I-HC)BQ_iC \\ I & & & \\ & \ddots & & \\ & & I & 0 \end{bmatrix}. \quad (9.17)$$

Similar with (9.3) and (9.7), $\bar{K}_2$ and $\bar{K}$ are given as below to eliminate the terms directly related to y_k in the estimation error dynamics:
$$\bar{K} = \bar{K}_1 + \bar{K}_2, \quad (9.18)$$
$$\bar{K}_2 = \bar{F}\bar{H}. \quad (9.19)$$

Unknown-Input-Observer Method

Considering the structures of $\bar{F}$, $\bar{K}$, and $\bar{H}$, the augmented estimation error $\bar{e}_k = \bar{x}_k - \hat{\bar{x}}_k$ can be partitioned as $\bar{e}_k = \left[e_k^T, \ldots, e_{k-l}^T\right]^T$. Based on the parameters obtained in (9.15)–(9.19), we have

$$e_{k+1} = Fe_k + (I - HC)BQ_iC\sum_{j=1}^{l} e_{k-j} - K_1 v_k + (I - HC)BQ_i \sum_{j=1}^{l} v_{k-j}$$
$$- Hv_{k+1} + (I - HC)w_k + (I - HC)f_k. \quad (9.20)$$

From (9.20), it is straightforward to see that the estimation error dynamics of the closed-loop system may be *unstable* even if F is Schur stable owing to the effects of the delayed errors $e_{k-1}, \ldots, e_{k-l}$. Such influences result from the integral part of the feedback control. Considering the sparse nature of $\tilde{C}$ in the closed-loop situation, the eigenvalues of $\bar{F}$ cannot be assigned effectively no matter how $\bar{K}_1$ is selected. To tackle such a problem, the structure of the closed-loop UIO should be appropriately changed.

Introducing the augmented measurement $\bar{y}_k = [y_k^T, y_{k-1}^T, \ldots, y_{k-l}^T]^T$, we have

$$\bar{y}_k = \tilde{C}\bar{x}_k + \bar{v}_k, \quad (9.21)$$

where $\tilde{C} = \text{diag}_l\{C\}$.

Now, consider a closed-loop UIO in the following form for system (9.12):

$$\begin{cases} \bar{z}_{k+1} = \bar{F}\bar{z}_k + \bar{K}\bar{y}_k, \\ \hat{\bar{x}}_{k+1} = \bar{z}_{k+1} + \bar{H}y_{k+1}, \end{cases} \quad (9.22)$$

where $\bar{z}_k \in \mathbb{R}^{nl}$ and $\hat{\bar{x}}_k \in \mathbb{R}^{nl}$ are the same as defined above. The modified UIO will be developed in the next theorem.

Theorem 9.1 *Consider the UIO (9.22) for the closed-loop system (9.1) with the PI controller (9.10). For the initial error $\bar{e}_0 = [e_0^T, 0, \ldots, 0]^T$, if $\bar{H}$ is given by (9.15), and $\bar{K}$ and $\bar{F}$ are calculated as follows:*

$$\bar{K} = \bar{K}_1 + \bar{K}_2, \quad (9.23)$$
$$\bar{F} = \bar{A} - \bar{H}\tilde{C}\bar{A} - \bar{K}_1\tilde{C}, \quad (9.24)$$
$$\bar{K}_1 = \begin{bmatrix} K_1 + (I - HC)BQ_p & (I - HC)BQ_i & \cdots & (I - HC)BQ_i \\ 0 & \cdots & \cdots & 0 \\ \vdots & \cdots & \cdots & \vdots \\ 0 & \cdots & \cdots & 0 \end{bmatrix}, \quad (9.25)$$
$$\bar{K}_2 = \bar{F}\bar{H}\bar{F}_2, \quad (9.26)$$

then the dynamics of the closed-loop residual $r_k = y_k - \bar{C}\hat{\bar{x}}_k$ will be identical with that in (9.8)–(9.9).

Proof Based on (9.12) and (9.22), we have

$$\begin{aligned}\bar{e}_{k+1} &= \bar{x}_{k+1} - (z_{k+1} + \bar{H}y_{k+1})\\ &= (I - \bar{H}\bar{C})\bar{x}_{k+1} - \bar{H}\bar{F}_2\bar{v}_k - \left[\bar{F}z_k + (\bar{K}_1 + \bar{K}_2)\bar{y}_k\right].\end{aligned} \quad (9.27)$$

Considering the definition of $\bar{y}_k$, we obtain that

$$\begin{aligned}\bar{e}_{k+1} &= (I - \bar{H}\bar{C})\bar{x}_{k+1} - \bar{H}\bar{F}_2\bar{v}_{k+1} - \bar{F}(\bar{x}_k - \bar{e}_k - \bar{H}y_k)\\ &\quad - \bar{K}_1(\tilde{C}\tilde{x}_k + \bar{v}_k) - \bar{K}_2\bar{y}_k.\end{aligned} \quad (9.28)$$

According to (9.12) and (9.24), (9.28) can be written as

$$\begin{aligned}\bar{e}_{k+1} &= \bar{F}\bar{e}_k - \bar{K}_1\bar{v}_k - \bar{H}\bar{F}_2\bar{v}_{k+1} + (I - \bar{H}\bar{C})\bar{E}d_k + (I - \bar{H}\bar{C})\bar{D}f_k\\ &\quad + (I - \bar{H}\bar{C})\bar{D}w_k + (I - \bar{H}\bar{C})\bar{F}_1\bar{v}_k + \bar{F}\bar{H}y_k - \bar{K}_2\bar{y}_k.\end{aligned} \quad (9.29)$$

From (9.15), (9.26) and the fact $y_k = \bar{F}_2\bar{y}_k$, it follows that

$$\begin{aligned}\bar{e}_{k+1} &= \bar{F}\bar{e}_k - \bar{K}_1\bar{v}_k - \bar{H}\bar{F}_2\bar{v}_{k+1} + (I - \bar{H}\bar{C})\bar{D}f_k + (I - \bar{H}\bar{C})\bar{D}w_k\\ &\quad + (I - \bar{H}\bar{C})\bar{F}_1\bar{v}_k.\end{aligned} \quad (9.30)$$

Now, we are in a position to investigate $\bar{F}$. With (9.24) and (9.25), we have

$$\begin{aligned}\bar{F} &= \bar{A} - \bar{H}\bar{C}\bar{A} - \bar{K}_1\tilde{C}\\ &= \begin{bmatrix} (I-HC)(A+BQ_pC) & (I-HC)BQ_iC & \cdots & (I-HC)BQ_iC \\ I & & & \\ & & \ddots & \\ & & I & 0 \end{bmatrix}\\ &\quad - \begin{bmatrix} K_1C + (I-HC)BQ_pC & (I-HC)BQ_iC & \cdots & (I-HC)BQ_iC \\ 0 & \cdots & \cdots & 0 \\ \vdots & \cdots & \cdots & \vdots \\ 0 & \cdots & \cdots & 0 \end{bmatrix}\end{aligned} \quad (9.31)$$

and it follows that

$$\bar{F} = \begin{bmatrix} F & 0 & \cdots & 0 \\ I & & & \\ & \ddots & & \\ & & I & 0 \end{bmatrix}. \quad (9.32)$$

With the proposed UIO parameters, the augmented estimation error $\bar{e}_k$ can still be partitioned as $\bar{e}_k = \left[e_k^T, \ldots, e_{k-l}^T\right]^T$. Taking $\bar{F}$ in (9.32) into account, we have

$$e_{k+1} = Fe_k - Hv_{k+1} - K_1v_k + (I - HC)w_k + (I - HC)f_k, \quad (9.33)$$

and

$$r_k = Ce_k + v_k, \quad (9.34)$$

which means that the dynamics of the closed-loop estimation error and the residual are exactly the same as those in (9.8) and (9.9) if the initial error $\bar{e}_0 = [e_0^T, 0, \ldots, 0]^T$. This concludes the proof.

Remark 9.1 *A modified UIO has been presented in Theorem 9.1 for a class of linear discrete-time system under PI control. The UIO has been developed with the control law, rather than the control input u_k. By doing so, the real-time signal transmission from the controller to the FDD unit can be avoided, and the relationship between the feedback control and the FDD performance can be established. At each time step, some historical measurements $y_{k-1}, y_{k-2}, \ldots, y_{k-l}$ are utilized to deal with the effects of the feedback control. It is noted that when only proportion control $u_k = Q_p y_k$ is employed, the UIO in (9.2) can be applied in the closed-loop case simply by setting that $\bar{K}_1 = K_1 + BQ_p - HCBQ_p$ and the dynamics of the closed-loop estimation error will be identical with that in (9.8) when the initial values are the same. In other words, the structure of the UIO is adjusted to specifically cope with the influences from the integral control. In the next section, a linear system without unknown disturbances will be considered, and both the Luenberger-observer-based method and the robust-observer-based method will be modified in the closed-loop case under PI control.*

9.2 Luenberger-Observer-Based and Robust-Observer-Based Methods

In this section, the Luenberger-observer-based and robust-observer-based methods will be investigated for a class of linear systems under PI control. Consider the following discrete-time system:

$$\begin{cases} x_{k+1} = Ax_k + Bu_k + f_k + w_k, \\ y_k = Cx_k + v_k, \end{cases} \quad (9.35)$$

where the variables x_k, u_k, y_k, f_k, w_k, and v_k and the same as defined in Section 9.1. The conventional Luenberger observer which directly uses the control input is of the following structure [109]:

$$\hat{x}_{k+1} = A\hat{x}_k + Bu_k + L(y_k - C\hat{x}_k), \quad (9.36)$$

where L is determined such that $A - LC$ is stable. The dynamics of the estimation error $e_k = x_k - \hat{x}_k$ and residual $r_k = y_k - C\hat{x}_k$ can be obtained, respectively, as follows:

$$e_{k+1} = (A - LC)e_k + w_k + f_k - Lv_k, \qquad (9.37)$$

$$r_k = Ce_k + v_k. \qquad (9.38)$$

Considering the same PI control (9.10), the system (9.35) can be written as

$$\begin{cases} \bar{x}_{k+1} = \bar{A}\bar{x}_k + \bar{D}f_k + \bar{D}w_k + \bar{F}_1\bar{v}_k, \\ y_k = \bar{C}\bar{x}_k + \bar{F}_2\bar{v}_k, \end{cases} \qquad (9.39)$$

where $\bar{x}_k$, $\bar{v}_k$, $\bar{A}$, $\bar{D}$, $\bar{F}_1$, $\bar{C}$, and $\bar{F}_2$ have been defined previously. For system (9.39), the following Luenberger observer is selected:

$$\hat{\bar{x}}_{k+1} = \bar{A}\hat{\bar{x}}_k + \bar{L}(y_k - \bar{C}\hat{\bar{x}}_k). \qquad (9.40)$$

Following the similar procedure in the previous section, if we choose

$$\bar{L} = [(L + BQ_p)^T, 0, \cdots, 0]^T, \qquad (9.41)$$

then

$$e_{k+1} = (A - LC)e_k + BQ_iC\sum_{j=1}^{l} e_{k-j} - Lv_k + BQ_i\sum_{j=1}^{l} v_{k-j} + w_k + f_k, \qquad (9.42)$$

from which we can conclude that the effects of the delayed errors may lead to unsatisfactory dynamics of the estimation error in the closed-loop system. To deal with this problem, we would resort to, once again, the augmented measurement output $\bar{y}_k$.

Consider a closed-loop Luenberger observer in the following form:

$$\hat{\bar{x}}_{k+1} = \bar{A}\hat{\bar{x}}_k + \bar{L}(\bar{y}_k - \tilde{C}\hat{\bar{x}}_k), \qquad (9.43)$$

where $\bar{L}$ is to be given in the next theorem.

Theorem 9.2 *Consider the Luenberger observer (9.43) for the closed-loop system (9.35) with the PI controller (9.10). If the initial error $\bar{e}_0 = [e_0^T, 0, \ldots, 0]^T$ and $\bar{L}$ is derived as:*

$$\bar{L} = \begin{bmatrix} L + BQ_p & BQ_i & \cdots & BQ_i \\ 0 & \cdots & \cdots & 0 \\ \vdots & \cdots & \cdots & \vdots \\ 0 & \cdots & \cdots & 0 \end{bmatrix}, \qquad (9.44)$$

then the dynamics of the closed-loop residual $r_k = y_k - \bar{C}\hat{\bar{x}}_k$ will be identical with that in (9.37)–(9.38).

Proof Noticing the fact that

$$\bar{A} - \bar{L}\tilde{C}$$

$$= \begin{bmatrix} A + BQ_pCBQ_iC \cdots BQ_iC \\ I \\ & \ddots \\ & & I & 0 \end{bmatrix} - \begin{bmatrix} LC + BQ_pCBQ_iC \cdots BQ_iC \\ 0 & \cdots & \cdots & 0 \\ \vdots & \cdots & \cdots & \vdots \\ 0 & \cdots & \cdots & 0 \end{bmatrix}$$

$$= \begin{bmatrix} A - LC & 0 & \cdots & 0 \\ I \\ & \ddots \\ & & I & 0 \end{bmatrix},$$

Theorem 9.2 can be easily obtained and the proof is omitted here for conciseness.

Now, let us discuss the robust observer presented in [47]. For system (9.35), the following observer is taken into account in the discrete situation:

$$\begin{cases} z_{k+1} = Fz_k + TBu_k + Gy_k, \\ r_{k+1} = Kz_{k+1} + Py_{k+1} \end{cases} \tag{9.45}$$

where

$$TA - FT = GC, \tag{9.46}$$
$$KT + PC = 0, \tag{9.47}$$
$$F \text{ is stable.} \tag{9.48}$$

For such an observer, r_k is the residual and the estimation error is defined as $e_k = z_k - Tx_k$. Based on (9.35) and (9.45), we have

$$e_{k+1} = Fe_k - Tw_k - Tf_k + Gv_k, \tag{9.49}$$

and

$$r_k = Ke_k + Pv_k. \tag{9.50}$$

For system (9.39), the following robust observer is selected:

$$\begin{cases} \bar{z}_{k+1} = \bar{F}\bar{z}_k + \bar{G}y_k, \\ r_{k+1} = \bar{K}\bar{z}_{k+1} + \bar{P}y_{k+1}, \end{cases} \tag{9.51}$$

With hope to make dynamics of the closed-loop error as similar as possible with that of the error obtained with control inputs, we choose the following parameters:

$$\bar{T} = \mathrm{diag}_l\{T\},\tag{9.52}$$

$$\bar{F} = \begin{bmatrix} F & 0 & \cdots & 0 \\ I & & & \\ & \ddots & & \\ & & I & 0 \end{bmatrix},\tag{9.53}$$

$$\bar{K} = [K, 0, \cdots, 0],\tag{9.54}$$

$$\bar{P} = P,\tag{9.55}$$

$$\bar{G} = [(G + TBQ_p)^T, 0, \cdots, 0]^T,\tag{9.56}$$

and then we have

$$e_{k+1} = Fe_k - TBQ_iC \sum_{j=1}^{l} e_{k-j} + Gv_k - TBQ_i \sum_{j=1}^{l} v_{k-j} - Tw_k - Tf_k.\tag{9.57}$$

Again, we need the augmented measurement to eliminate the influences of the integral control. For this purpose, we consider a closed-loop robust observer in the following form for system (9.35):

$$\begin{cases} \bar{z}_{k+1} = \bar{F}\bar{z}_k + \bar{G}\bar{y}_k, \\ r_{k+1} = \bar{K}\bar{z}_{k+1} + \bar{P}y_{k+1}. \end{cases}\tag{9.58}$$

In virtue of Theorem 9.1, the following results are easily accessible.

Theorem 9.3 *Consider the robust observer (9.58) for the closed-loop system (9.35) with the PI controller (9.10). For the initial error $\bar{e}_0 = [e_0^T, 0, \ldots, 0]^T$, if $\bar{T}, \bar{F}, \bar{K},$ and $\bar{P}$ are given by (9.52)–(9.55) and $\bar{G}$ is adopted as:*

$$\bar{G} = \begin{bmatrix} G + TBQ_p & TBQ_i & \cdots & TBQ_i \\ 0 & \cdots & \cdots & 0 \\ \vdots & \cdots & \cdots & \vdots \\ 0 & \cdots & \cdots & 0 \end{bmatrix},\tag{9.59}$$

then the dynamics of the closed-loop residual r_k will be identical with that in (9.49)–(9.50).

Proof The proof of Theorem 9.3 is similar with that of Theorem 9.1 and is therefore omitted.

Remark 9.2 *So far, several observer-based FDD methods (including UIO-based method, Luenberger-observer-based method, and robust-observer-based method) have been examined in the closed-loop situations. When only control laws are available, the effects of the integral control have been analyzed and they* may degrade the state estimation performances. Fortunately, by appropriately changing the structure of the observers, the dynamics of the closed-loop residuals can be made exactly the same as that of residuals obtained with known control inputs. The scalar l and the matrices Q_p and Q_i quantify the effects of the controller on the dimensions and parameterizations of the observers, respectively. According to Theorems 9.1–9.3, the parameters of the closed-loop observers can be constructed with the open-loop observer parameters and the original system parameters after several matrix additions and matrix multiplications. The computation burden of the presented method would not get significantly increased even with higher parameter dimensions, which makes the method easy to implement in practice. It is also worth mentioning that the results obtained in the chapter can be easily extended to time-varying systems by using Theorems 9.1–9.3 at each time step. For uncertain systems in which the directions of the uncertainties are known, the UIO-based FDD method can be utilized after formulating the uncertainties as parts of unknown inputs. When the directions of the uncertainties are unavailable, the H_∞ observer can be used to constrain the attenuation level from the uncertainties to the residuals. The structure of the observer should be modified, and the parameters should be redesigned accordingly. For the systems under the network-based environments, the network-induced phenomena such as time delays and packet dropouts [131, 162] will lead to that the influences of the possible faults on the residuals are stochastic and the dynamics of residuals is directly dependent on the stochastic variables and the original system states. In such a case, it makes more sense to study the FDD problem and adjust the observer parameters in the robust or mean-square sense, and the residual evaluation problem becomes more challenging. In the next section, an illustrative example will be provided to show the effectiveness of the proposed closed-loop algorithm in the presence of the PI controller.

9.3 A Simulation Example

Inspired by the three-tank system proposed in [62], a system (9.1) with the following parameters is considered:

$$A = \begin{bmatrix} 0.9908 & 0 & 0.0091 \\ 0 & 0.9856 & 0.0072 \\ 0.0091 & 0.0072 & 0.9836 \end{bmatrix}, \ B = \begin{bmatrix} 64.6627 & 0.0007 & 0.2978 \\ 0.0007 & 64.4908 & 0.2358 \\ 0.2978 & 0.2358 & 64.4217 \end{bmatrix}, \ E = \begin{bmatrix} 2 \\ 1 \\ 0 \end{bmatrix},$$

$$C = \begin{bmatrix} 1 & 0 & 0 \\ 0 & 1 & 0 \end{bmatrix}.$$

The closed-loop control is designed as $u_k = Q_p y_k + Q_i \sum_{i=1}^{2} y_{k-i}$ where

$$Q_p = \begin{bmatrix} -0.0029 & 0.0087 \\ 0.0001 & -0.0057 \\ -0.0372 & -1.8862 \end{bmatrix}, \quad Q_i = \begin{bmatrix} 0 & 0.0001 \\ -0.004 & 0 \\ 0.001 & 0.001 \end{bmatrix}.$$

The disturbances d_k, w_k, and v_k are mutually independent Gaussian distributed sequences whose variances are $1 \times 10^{-10} I$, $1 \times 10^{-12} I$ and $1 \times 10^{-12} I$, respectively. The observer parameters in (9.3)–(9.7) are determined as

$$H = \begin{bmatrix} 0 & 2 \\ 0 & 1 \\ 0 & 0 \end{bmatrix}, K_1 = \begin{bmatrix} 0.4 & 0 \\ 0 & 0.3 \\ 0 & 0 \end{bmatrix}, \quad K_2 = \begin{bmatrix} 0 & -0.7896 \\ 0 & -0.3 \\ 0 & 0.0254 \end{bmatrix},$$

$$F = \begin{bmatrix} 0.5908 & -1.9712 & -0.0053 \\ 0 & -0.3 & 0 \\ 0.0091 & 0.0072 & 0.9836 \end{bmatrix}.$$

In such a case, F is stable and its eigenvalues are 0.5909, 0.9835, and -0.3.
The additive fault f_k is in the following form:

$$f_k = \begin{cases} [0, 0, 0]^T, & \text{if } k \le 20, \\ [-8 \times 10^{-6}, 0, 0]^T, & \text{otherwise.} \end{cases}$$

In the fault-free case, the average Euclidean norm of the residual obtained with the proposed closed-loop observer is 1.9983×10^{-4} in 50 Monte-Carlo simulations. With the traditional open-loop observer, the average Euclidean norm of the residual is 4.4595×10^{-4}. Fig. 9.1 plots the norm of residual achieved with the conventional UIO method (9.13) in the fault-free case. It can be seen that: 1) the closed-loop UIO can achieve better estimation performance than the open-loop one due to the consideration of the feedback control; 2) the residual obtained with the open-loop observer cannot converge even when there is no fault in the system. As a result, the conventional UIO is no longer applicable in the closed-loop case. Fig. 9.2 depicts the norm of residual obtained with the closed-loop UIO method (9.22) in the faulty situation and the threshold obtained from 100 Monte-Carlo experiments. We can conclude that the closed-loop UIO method proposed in (9.22) could help us to detect the fault effectively thanks to our efforts to deal with the closed-loop system dynamics.

A Simulation Example

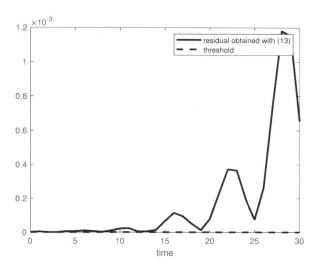

FIGURE 9.1: Residual obtained with (9.13)

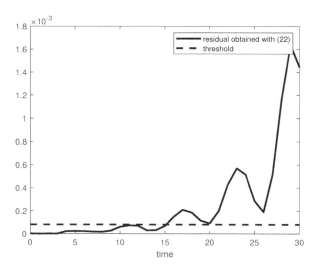

FIGURE 9.2: Residual obtained with (9.22)

9.4 Conclusion

In this chapter, the FDD problems under closed-loop control have been investigated with a class of observer-based methods. A discrete-time linear system with/without unknown inputs under PI control has been considered. Several observer-based FDD methods have been taken into account including UIO method, Luenberger observer method, and robust observer method. It has been shown that the approaches may be no longer applicable in closed-loop situations when only control laws are available due to the integral feedback control. To solve the problem, the structures of the observers have been modified and the parameters have been adjusted such that the dynamics of closed-loop residuals are identical with that of the residuals obtained with known control inputs. A numerical example has been presented to show the effectiveness of the proposed algorithm. Future research topics would include the extension of our results to systems with more complex controls (e.g. sliding-mode control and adaptive control), and more complicated practical systems such as uncertain systems, nonlinear systems, and networked systems.

10

State Estimation and Fault Reconstruction with Integral Measurements under Partially Decoupled Disturbances

Over the past few decades, modern industrial systems have become more and more complex at an exponentially increasing expense, and such a trend has led to higher and higher safety/reliability demands. Consequently, the research on the fault diagnosis (FD) problem has been gaining momentum and considerable effort has been devoted to the detection of occurrence of faults, the identification of the position of faults as well as the determination of the amplitude of faults. So far, the FD problem has been extensively studied in many practical systems including induction motors, robotic systems, wind turbines, power systems, planetary gearboxes, three-phase inverters, flight control systems, and so forth. Among existing FD approaches, model-based FD methods have been widely adopted and a great number of results have been available in the literature on model-based FD issues.

In engineering practice, the FD performance indices, such as the false alarm rate and the missing alarm rate, usually degrade by ubiquitous additive disturbances. Naturally, an important task in FD research is how to eliminate, or at least mitigate the impacts from the disturbances onto the diagnosis performance. To this end, a widely employed approach is to make sure the insensitivity of FD procedure against the disturbances. Based on this basic idea, the H_-/H_∞ criterion has been widely adopted to calculate the residual signals, which are meant to be sensitive to the faults and insensitive to the disturbances. In addition, the error bound on the worst-case fault estimation with respect to the disturbance has been investigated in the H_∞ framework when the fault needs to be directly estimated. An alternative approach is to decouple the disturbances from the residual so that the effects of the disturbances on the FD results can be completely eliminated. Since proposed in 1996 [23], the UIO technique has been frequently used to decompose the disturbances from the estimation errors and/or residuals. Under certain assumptions on the measurement matrix and the disturbance distribution matrix, the desired decoupling relationship can be realized in the UIO.

The UIO-based FD strategies have been applied to many situations such as nonlinear systems, fuzzy systems, singular systems, multi-agent systems, etc. It is worth mentioning that only under some rather stringent conditions, the traditional UIO can achieve a complete decoupling from the disturbances.

In reality, however, it is often the case that the disturbances can only be *partially decoupled* and some parts of the disturbances are *undecouplable*. So far, some relevant results have been reported in the literature. For example, in [159], the state estimation problem has been discussed for systems subject to partially decoupled disturbances. The fault estimation problem for systems with partially decoupled disturbance has been elegantly solved in [44] under the assumption that the qth-order derivative of the fault is zero. Nevertheless, there is still much room for further development, for example, the examination on the FD problem for *a wider range of faults* with partially decoupled disturbances, and this constitutes one of our motivations.

In most FD studies, the measurement outputs have been implicitly assumed to be dependent on the system state at the *current* time step. This assumption is, unfortunately, not always true in practical engineering. In reality, the system measurements might be in proportion to the integral of the system states over a given time period due probably to the delayed data collection and the real-time signal processing. This phenomenon, known as *integral measurement*, often occurs in engineering applications such as chemical processes and nuclear reaction processes. In [54], a novel model has been put forward to describe the integral measurement and a modified unscented Kalman filter has been developed to estimate the states of the addressed nonlinear system.

Despite its clear practical relevance in process engineering, the fault estimation problem for systems with integral measurements has not received adequate research attention yet, not to mention the situation where the partially decoupled disturbances are also taken into account. As such, a natural research topic of both theoretical and practical importance is to study the state estimation and fault reconstruction problems for systems with integral measurements under partially decoupled disturbances. This issue appears to be non-trivial due primarily to the following identified challenges: 1) how to describe the partially decoupled disturbances and integral measurements within a unified framework? 2) how to develop a feasible design algorithm for the observer that can decouple the disturbances as much as possible and attenuate the effects of the undecouplable disturbances? and 3) how to establish the existence conditions for the desired observer? It is, therefore, the main purpose of this chapter to answer these three questions.

In this chapter, the state estimation and fault reconstruction problems are investigated for a class of systems with integral measurements under partially decoupled disturbances. An augmented state is constructed to combine the current system state, the delayed system state, and the possible additive fault. Then, the addressed system is written as a singular system. By doing so, we no longer need to assume that the qth-order derivative of the fault is zero or the dynamics of the fault is available to us [44]. The existence conditions for the UIO decoupling partial disturbances are established and the observer gains are calculated by resorting to the LMI technique, where the impacts from the undecouplable disturbances on the state estimation and the fault reconstruction are attenuated, and the decouplable disturbances are decoupled

Problem Formulation 165

as much as possible. To show the effectiveness of the presented method, a simulation example based on a three-tank system is provided.

The main contributions of the chapter can be highlighted as follows: *1) a comprehensive model is considered which accounts for both partially decoupled disturbances and integral measurements; 2) a novel UIO is designed that decouples the disturbances as much as possible and also attenuates the effects of the undecouplable disturbances on the accuracies of the state estimation and the fault reconstruction; and 3) the explicit design procedures and existence conditions for the desired observer are established.*

10.1 Problem Formulation

Consider the following class of discrete-time systems:

$$\begin{cases} x_{k+1} = Ax_k + Bu_k + Dd_k + Ff_k, \\ y_k = C\sum_{i=0}^{s} x_{k-i} + Gf_k + Hv_k, \end{cases} \quad (10.1)$$

where $x_k \in \mathbf{R}^n$ is the system state; $u_k \in \mathbf{R}^b$ is the control input; $y_k \in \mathbf{R}^m$ is the measurement output; $f_k \in \mathbf{R}^l$ is the considered additive fault; $d_k \in \mathbf{R}^q$ is the unknown input stemming possibly from disturbances, modelling errors and/or some other possible phenomena; $v_k \in \mathbf{R}^p$ is the measurement noise. s is the time interval to collect the data. A, B, C, D, F, G, and H are known matrices with appropriate dimensions. Moreover, d_k and D can be partitioned as $d_k = [d_{1,k}^T, d_{2,k}^T]^T$ and $D = [D_1, D_2]$, where $d_{1,k} \in \mathbf{R}^{q_1}$ and $d_{2,k} \in \mathbf{R}^{q_2}$. $d_{1,k}$ is the unknown input that can be decoupled, and D_1 is assumed to be of full column rank. $d_{2,k}$ and v_k are assumed to be square-summable.

Define an augmented state as $\eta_k = [x_k^T, \ldots, x_{k-s}^T, f_k^T]^T$. Then, system (10.1) can be re-written in the following singular form:

$$\begin{cases} E\eta_{k+1} = \bar{A}\eta_k + \bar{B}u_k + \bar{D}d_k, \\ y_k = \bar{C}\eta_k + Hv_k, \end{cases} \quad (10.2)$$

where

$$\bar{A} = \begin{bmatrix} A & 0 & \cdots & 0 & F \\ I & 0 & \cdots\cdots & 0 \\ 0 & \ddots & \cdots\cdots & \vdots \\ \vdots & \cdots & \ddots & \cdots & \vdots \\ 0 & \cdots\cdots & I & 0 \end{bmatrix} \in \mathbf{R}^{(s+1)n \times [(s+1)n+l]},$$

$$\bar{B} = \begin{bmatrix} B \\ 0 \end{bmatrix} \in \mathbf{R}^{(s+1)n \times b}, \quad \bar{D} = \begin{bmatrix} D \\ 0 \end{bmatrix} \in \mathbf{R}^{(s+1)n \times q},$$

$$E = [I, 0] \in \mathbf{R}^{(s+1)n \times [(s+1)n+l]}, \quad \bar{C} = [C, \cdots, C, G] \in \mathbf{R}^{m \times [(s+1)n+l]}.$$

Similar with D, $\bar{D}$ can be partitioned as $\bar{D} = [\bar{D}_1, \bar{D}_2]$, where

$$\bar{D}_1 = \begin{bmatrix} D_1 \\ 0 \end{bmatrix}, \quad \bar{D}_2 = \begin{bmatrix} D_2 \\ 0 \end{bmatrix}.$$

To facilitate the feasibility study of the UIO, the following assumption is made.

$$\operatorname{rank} \begin{bmatrix} E \\ \bar{C} \end{bmatrix} = (s+1)n + l. \tag{10.3}$$

In this chapter, the observer for system (10.2) is of the following structure:

$$\begin{cases} z_{k+1} = M z_k + N u_k + J y_k, \\ \hat{\eta}_{k+1} = z_{k+1} + L y_{k+1}, \end{cases} \tag{10.4}$$

where $z_k \in \mathbf{R}^{(s+1)n+l}$ is the observer state and $\hat{\eta}_k \in \mathbf{R}^{(s+1)n+l}$ is the estimate of η_k. The matrices M, N, J, and L are observer gains to be designed.

Denote $e_k = \eta_k - \hat{\eta}_k$ and $w_{k+1} = \begin{bmatrix} d_{2,k}^T, v_{k+1}^T \end{bmatrix}^T$. The goal of the addressed problem is to design an observer (10.4) for system (10.2), where the following requirements can be satisfied:

1. The estimation error e_{k+1} can be fully decoupled from $d_{1,k}$.
2. Under the zero-initial condition, the estimation error e_k satisfies

$$\sum_{k=0}^{\infty} \|e_k\|^2 < \gamma^2 \sum_{k=1}^{\infty} \|w_k\|^2 \tag{10.5}$$

for all nonzero w_k, where $\gamma > 0$ is a predefined attenuation level.

Remark 10.1 *The considered fault may result from some state-dependent changes such as actuator or sensor failures in a control system. It is worth mentioning that the system (10.1) is comprehensive since both partially decoupled disturbance and integral measurements are included in a unified framework. In practice, it is quite common that only partial disturbance can be decoupled from the procedure of state estimation and fault reconstruction. As such, to mitigate the side effects of the disturbances, there is a practical need to decouple the disturbances as much as possible and, at the same time, attenuate the undecouplable disturbances. The full column rank of D_1 can be ensured by appropriately constructing the decouplable unknown input $d_{1,k}$. Furthermore, delayed data collection and signal processing may lead to the integral measurements. In an extreme situation of $s = 0$, the measurement equation will reduce to the usual case where the measurement outputs are only dependent on the current system state. The state estimation and fault reconstruction problems with integral measurements under partially decoupled disturbances have not yet been addressed in the literature, and this gives rise to the main motivation of our work.*

10.2 Filter Design

In this section, we establish the existence conditions of the desired observer, that is, the conditions under which e_{k+1} can be decoupled from $d_{1,k}$.

Theorem 10.1 *Consider the UIO (10.4) for the singular system (10.2). If there exists a matrix $X \in \mathbf{R}^{[(s+1)n+l] \times (s+1)n}$ such that the following equations hold:*

$$XE + L\bar{C} = I, \qquad (10.6)$$
$$X\bar{D}_1 = 0, \qquad (10.7)$$
$$X\bar{B} = N, \qquad (10.8)$$
$$X\bar{A} = M, \qquad (10.9)$$
$$ML = J, \qquad (10.10)$$

then the dynamics of the estimation error e_{k+1} can be fully decoupled from $d_{1,k}$.

Proof Based on (10.1) and (10.4), we have

$$\begin{aligned} e_{k+1} &= \eta_{k+1} - \hat{\eta}_{k+1} \\ &= \eta_{k+1} - z_{k+1} - L(\bar{C}\eta_{k+1} + Hv_{k+1}). \end{aligned} \qquad (10.11)$$

Considering (10.4) and (10.6), we obtain that

$$e_{k+1} = XE\eta_{k+1} - Mz_k - Nu_k - Jy_k - LHv_{k+1}. \qquad (10.12)$$

According to (10.2), (10.12) can be written as

$$e_{k+1} = X\bar{A}\eta_k + X\bar{B}u_k + X\bar{D}d_k - Mz_k - Nu_k - Jy_k - LHv_{k+1}. \qquad (10.13)$$

Substituting (10.7)-(10.9) into (10.13) yields

$$e_{k+1} = M\eta_k + X\bar{D}_2 d_{2,k} - Mz_k - Jy_k - LHv_{k+1}. \qquad (10.14)$$

From (10.4), it follows that

$$e_{k+1} = M\eta_k + X\bar{D}_2 d_{2,k} - M(\eta_k - e_k - Ly_k) - Jy_k - LHv_{k+1}. \qquad (10.15)$$

Then, based on (10.10), we have

$$e_{k+1} = Me_k + X\bar{D}_2 d_{2,k} - LHv_{k+1}, \qquad (10.16)$$

which is independent of $d_{1,k}$, and the proof is complete.

Denote $\Lambda = \begin{bmatrix} X\bar{D}_2, -LH \end{bmatrix}$. According to the definition of w_k, (10.16) can be written as:

$$e_{k+1} = Me_k + \Lambda w_{k+1}. \qquad (10.17)$$

Next, the disturbance attenuation index (10.5) will be considered. In view of (10.17), sufficient conditions will be established in the following theorem in order to ensure the requirement (10.5).

Theorem 10.2 *If there exist a matrix X satisfying (10.6)–(10.10) and a positive definite matrix $P = P^T > 0$ such that the following inequality*

$$\begin{bmatrix} -P & 0 & 0 & \bar{A}^T X^T P \\ * & -\gamma^2 I & 0 & \bar{D}_2^T X^T P \\ * & * & -\gamma^2 I & -H^T L^T P \\ * & * & * & -P \end{bmatrix} < 0 \qquad (10.18)$$

holds, then the performance index (10.5) is met.

Proof Choosing the Lyapunov function $V_k = e_k^T P e_k$, it follows readily from (10.17) that

$$\begin{aligned} V_{k+1} - V_k = & w_{k+1}^T \Lambda^T P \Lambda w_{k+1} + w_{k+1}^T \Lambda^T P M e_k \\ & + e_k M^T P \Lambda w_{k+1} + e_k^T (M P M^T - P) e_k. \end{aligned} \qquad (10.19)$$

Then, we have

$$\begin{aligned} & \sum_{k=0}^{N} \left(e_k^T e_k - \gamma^2 w_{k+1}^T w_{k+1} \right) \\ = & \sum_{k=0}^{N} \left(e_k^T e_k - \gamma^2 w_{k+1}^T w_{k+1} + V_{k+1} - V_k \right) + V_0 - V_{N+1} \\ = & \sum_{k=0}^{N} \xi_{k+1}^T \Omega \xi_{k+1} + V_0 - V_{N+1}, \end{aligned} \qquad (10.20)$$

where $\xi_{k+1} = \begin{bmatrix} e_k^T, w_{k+1}^T \end{bmatrix}^T$ and

$$\Omega = \begin{bmatrix} M P M^T - P & M^T P \Lambda \\ * & -\gamma^2 I + \Lambda^T P \Lambda \end{bmatrix}. \qquad (10.21)$$

Considering $V_{N+1} > 0$ and $V_0 = 0$ under the zero-initial condition, and letting $N \to \infty$, we obtain

$$\sum_{k=0}^{\infty} \left(e_k^T e_k - \gamma^2 w_{k+1}^T w_{k+1} \right) < \sum_{k=0}^{\infty} \xi_{k+1}^T \Omega \xi_{k+1}. \qquad (10.22)$$

As a result, the requirement (10.5) is guaranteed as long as $\Omega < 0$. By resorting to the Schur complement, $\Omega < 0$ holds if and only if the following is true:

$$\begin{bmatrix} -P & 0 & M^T P \\ * & -\gamma^2 I & \Lambda^T P \\ * & * & -P \end{bmatrix} < 0. \qquad (10.23)$$

Based on (10.9) and the definition of Λ, (10.18) holds if and only if (10.23) holds, and the proof is complete.

Parameter Calculation 169

Remark 10.2 *Assumption (10.3) can guarantee the solvability of (10.6). The constraints (10.6) and (10.7) are put forward to guarantee that the influences of $d_{1,k}$ on the estimation errors can be fully eliminated. It is noted that the parameters in (10.18) cannot be determined under the equality constraints (10.6)–(10.10). In the next section, a feasible way to calculate the observer gains will be provided such that the two indices proposed in Section 10.1 can be satisfied simultaneously.*

10.3 Parameter Calculation

Before proceeding further, rewrite the equalities (10.6) and (10.7) as:

$$[X\ L] \begin{bmatrix} E\bar{D}_1 \\ \bar{C} & 0 \end{bmatrix} = [I\ 0]. \tag{10.24}$$

As discussed in [84], (10.24) is solvable if the following condition holds:

$$\operatorname{rank} \begin{bmatrix} E\bar{D}_1 \\ \bar{C} & 0 \end{bmatrix} = (s+1)n + l + \operatorname{rank}(\bar{D}_1). \tag{10.25}$$

When (10.25) is met, the solution of (10.24) can be constructed as follows:

$$[X\ L] = [I\ 0] \begin{bmatrix} E\bar{D}_1 \\ \bar{C} & 0 \end{bmatrix}^+ + K \left\{ I - \begin{bmatrix} E\bar{D}_1 \\ \bar{C} & 0 \end{bmatrix} \begin{bmatrix} E\bar{D}_1 \\ \bar{C} & 0 \end{bmatrix}^+ \right\}, \tag{10.26}$$

where $K \in \mathbf{R}^{[(s+1)n+l]\times[(s+1)n+m]}$ provides extra degree of freedom for determining the desired observer gains.

Define

$$S_1 = [I\ 0] \begin{bmatrix} E\bar{D}_1 \\ \bar{C} & 0 \end{bmatrix}^+ \begin{bmatrix} I \\ 0 \end{bmatrix}, \tag{10.27}$$

$$S_2 = [I\ 0] \begin{bmatrix} E\bar{D}_1 \\ \bar{C} & 0 \end{bmatrix}^+ \begin{bmatrix} 0 \\ I \end{bmatrix}, \tag{10.28}$$

$$S_3 = \left\{ I - \begin{bmatrix} E\bar{D}_1 \\ \bar{C} & 0 \end{bmatrix} \begin{bmatrix} E\bar{D}_1 \\ \bar{C} & 0 \end{bmatrix}^+ \right\} \begin{bmatrix} I \\ 0 \end{bmatrix}, \tag{10.29}$$

$$S_4 = \left\{ I - \begin{bmatrix} E\bar{D}_1 \\ \bar{C} & 0 \end{bmatrix} \begin{bmatrix} E\bar{D}_1 \\ \bar{C} & 0 \end{bmatrix}^+ \right\} \begin{bmatrix} 0 \\ I \end{bmatrix}. \tag{10.30}$$

Then, according to (10.26), we have

$$X = S_1 + KS_3, \tag{10.31}$$
$$L = S_2 + KS_4. \tag{10.32}$$

After calculating N, M, and J according to (10.8)–(10.10) with X and L given in (10.31) and (10.32), respectively, the desired decoupling relationship in Theorem 10.1 can be established, and it remains to select the matrix K such that the robustness criterion in Theorem 10.2 can be met. The following theorem is to be developed with hope to appropriately determine K.

Theorem 10.3 *If there exist a matrix $Y \in \mathbf{R}^{[(s+1)n+m] \times [(s+1)n+l]}$ and a positive definite matrix $P = P^T > 0$ satisfying the following inequality:*

$$\begin{bmatrix} -P & 0 & 0 & \bar{A}^T S_1^T P + \bar{A}^T S_3^T Y \\ * & -\gamma^2 I & 0 & \bar{D}_2 S_1^T P + \bar{D}_2 S_3^T Y \\ * & * & -\gamma^2 I & -H^T S_2^T P - H^T S_4^T Y \\ * & * & * & -P \end{bmatrix} < 0, \qquad (10.33)$$

then both the disturbance decoupling and the disturbance attenuation requirements are guaranteed with

$$K = P^{-1} Y^T. \qquad (10.34)$$

The parameters X and L can be calculated with (10.31) and (10.32), and N, M and J can then be determined with (10.8), (10.9), and (10.10), respectively.

Proof *Substituting (10.31) and (10.32) into (10.18) yields*

$$\begin{bmatrix} -P & 0 & 0 & \bar{A}^T S_1^T P + \bar{A}^T S_3^T K^T P \\ * & -\gamma^2 I & 0 & \bar{D}_2 S_1^T P + \bar{D}_2 S_3^T K^T P \\ * & * & -\gamma^2 I & -H^T S_2^T P - H^T S_4^T K^T P \\ * & * & * & -P \end{bmatrix} < 0. \qquad (10.35)$$

By setting $Y = K^T P$, (10.33) follows directly from (10.35). Considering the conclusions in Theorems 10.1–10.2, Theorem 10.3 can be obtained.

Remark 10.3 *Theorem 10.3 presents a practical way to determine the observer gains with hope to simultaneously satisfy the decoupling and attenuation requirements against the disturbances. By introducing the variable K, the linear constraints in (10.6) and (10.7) can be integrated into a linear matrix inequality, which facilitates the feasibility of the UIO design problem, and the disturbances are therefore decoupled as much as possible and the effects of the undecouplable disturbances are mitigated as well. Furthermore, the existence conditions for the desired observer gains are explicitly characterized. The distribution matrices constitute the main difference between the disturbances and additive faults, and the difference offers room for attenuating the effects of disturbances and reconstructing the faults simultaneously. The proposed method can deal with the case that the augmented disturbance w_{k+1} is square-summable and the distribution matrix of the decouplable disturbance $d_{1,k}$ is of full rank and satisfies (10.25). The consideration of partially decoupled disturbances and integral measurements, the singular structure of the*

Illustrative Example 171

augmented system, and the novel observer requirement and design procedure constitute the main differences between our work and existing methods. In the next section, a numerical example is used to demonstrate the effectiveness of the proposed method on the state estimation and fault reconstruction problems in the presence of integral measurements and partially decoupled disturbances.

10.4 Illustrative Example

Inspired by the three-tank system in [58], we consider the system (10.1) with the following parameters:

$$A = \begin{bmatrix} 0.9908 & 0 & 0.0091 \\ 0 & 0.9856 & 0.0072 \\ 0.0091 & 0.0072 & 0.9836 \end{bmatrix}, \quad B = \begin{bmatrix} 64.6627 & 0.0007 & 0.2978 \\ 0.0007 & 64.4908 & 0.2358 \\ 0.2978 & 0.2358 & 64.4217 \end{bmatrix},$$

$$D = \begin{bmatrix} 1 & 0 & 0 \\ 0 & 1 & 0 \\ 0 & 0 & 1 \end{bmatrix}, \quad F = \begin{bmatrix} 1 \\ 0 \\ 0 \end{bmatrix}, \quad C = \begin{bmatrix} 1 & 0 & 0 \\ 0 & 1 & 0 \end{bmatrix}, \quad G = \begin{bmatrix} 1 \\ 1 \end{bmatrix}, \quad H = \begin{bmatrix} 1 & 0 \\ 0 & 1 \end{bmatrix}.$$

The closed-loop control is designed as $u_k = Q_p y_k$, where

$$Q_p = 10^{-3} \times \begin{bmatrix} 0.0435 & -0.1305 \\ -0.0021 & 0.0855 \\ 0.5577 & 28.2936 \end{bmatrix}.$$

The disturbances d_k and v_k obey that

$$d_k = 10^{-6} \times e^{-3k} [1, 2, 3]^T,$$
$$v_k = 4 \times 10^{-6} \times e^{-0.1k} [\sin k, \sin 2k]^T.$$

The time interval to collect the data s is set to be 2, the attenuation level γ is $\sqrt{5}$, and the first component of d_k can be decoupled from the observer, i.e.,

$$D_1 = \begin{bmatrix} 1 \\ 0 \\ 0 \end{bmatrix}, \quad D_2 = \begin{bmatrix} 0 & 0 \\ 1 & 0 \\ 0 & 1 \end{bmatrix}.$$

Such a choice can guarantee that the assumption (10.25) is satisfied. In this system, $x_k \in \mathbf{R}^3$ represents the liquid levels of the three tanks; $u_k \in \mathbf{R}^2$ is the liquid inflow; $y_k \in \mathbf{R}^2$ is the measurement output denoting the liquid heights of tank 1 and tank 2; $d_k \in \mathbf{R}^3$ is the liquid height disturbance; $f_k \in \mathbf{R}$ is the considered fault signal. With the given parameters, the LMI in Theorem 10.3 can be solved and, according to (10.31), (10.32), and (10.8)-(10.10), the observer gains are obtained as follows:

$$M = \begin{bmatrix} -0.9998 & 1.9852 & 0.0073 & -0.9999 & 0.9998 & 0 & 0 & 0 & 0 & 0 \\ 0 & 0.9856 & 0.0072 & 0 & 0 & 0 & 0 & 0 & 0 & 0 \\ 0.0091 & 0.0072 & 0.9836 & 0 & 0 & 0 & 0 & 0 & 0 & 0 \\ 0.9995 & 0.0002 & -0.0005 & 0.0002 & 0 & 0 & 0 & 0 & 0 & 0 \\ 0 & 1 & 0 & 0 & 0 & 0 & 0 & 0 & 0 & 0 \\ 0 & 0 & 1 & 0 & 0 & 0 & 0 & 0 & 0 & 0 \\ 0.0003 & 0.0002 & 0.0004 & 0.9996 & 0.0001 & 0 & 0 & 0 & 0 & 0 \\ 0 & 0 & 0 & 0 & 1 & 0 & 0 & 0 & 0 & 0 \\ 0 & 0 & 0 & 0 & 0 & 1 & 0 & 0 & 0 & 0 \\ 0 & -1.9856 & -0.0072 & 0 & -1 & 0 & 0 & 0 & 0 & 0 \end{bmatrix}$$

$$N = \begin{bmatrix} 0.0007 & 64.4782 & 0.2396 \\ 0.0007 & 64.4906 & 0.2357 \\ 0.2978 & 0.2357 & 64.4269 \\ -0.0028 & 0.0021 & -0.0309 \\ 0 & -0.0002 & -0.0001 \\ 0.0001 & -0.0001 & 0.0001 \\ 0.0029 & 0.0100 & 0.0268 \\ 0 & -0.0001 & -0.0001 \\ 0 & -0.0002 & -0.0002 \\ -0.0008 & -64.4903 & -0.2355 \end{bmatrix},$$

$$J = \begin{bmatrix} -0.9997 & 0.9997 \\ 0 & 0 \\ 0.0091 & -0.0091 \\ 0.9993 & -0.9994 \\ 0 & 0 \\ 0 & 0 \\ 0.0004 & -0.0003 \\ 0 & 0 \\ 0 & 0 \\ 0 & 0 \end{bmatrix}, L = \begin{bmatrix} 0.9998 & -0.9999 \\ 0 & 0 \\ 0 & 0 \\ 0.0001 & 0 \\ 0 & 0 \\ 0 & 0 \\ 0.0001 & -0.0001 \\ 0 & 0 \\ 0 & 0 \\ 0 & 1.0000 \end{bmatrix}.$$

The considered additive fault f_k is in the following form:

$$f_k = \begin{cases} 0, & \text{if } k \leq 26, \\ -3(k-26) \times 10^{-6}, & \text{otherwise.} \end{cases}$$

Figs. 10.1–10.3 show the actual states and their estimates obtained with the developed observer, respectively, and Fig. 10.4 illustrates the actual additive fault and its estimate. It can be concluded that the proposed observer can estimate the system states and the additive fault well due to our effort in handling the integral measurements and partially decoupled disturbances. The

Illustrative Example

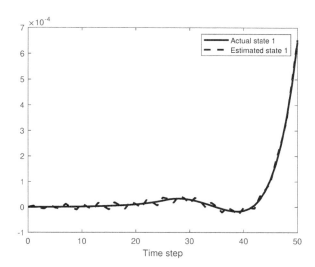

FIGURE 10.1: The state x_1 and its estimate

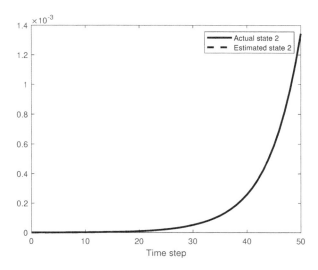

FIGURE 10.2: The state x_2 and its estimate

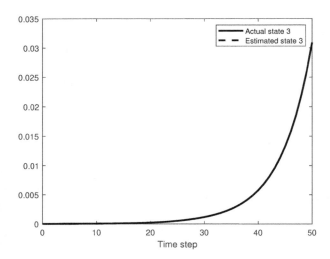

FIGURE 10.3: The state x_3 and its estimate

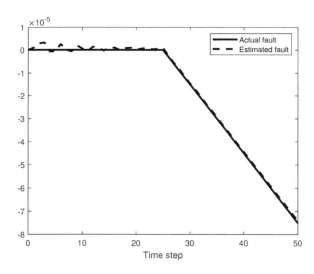

FIGURE 10.4: The actual additive fault and its estimate

average norm of the fault estimation error obtained with the proposed method is 1.5258×10^{-7}, and that obtained with the traditional disturbance decomposition approach without considering the H_∞ requirement is 9.8674×10^{-7}. So the proposed observer can better solve the fault reconstruction problem. The magnitude orders of the disturbances and faults have been set according to the practical values in real three-tank systems.

10.5 Conclusion

The state estimation and fault reconstruction problems have been addressed in the chapter for a class of discrete systems with integral measurements under partially decoupled disturbances. An augmented state consisting of the current system state, the delayed system state, and the additive fault has been considered and then a singular system has been established. The UIO has been obtained for the singular system such that the disturbances can be decoupled as much as possible and the effects of the disturbances that cannot be decoupled have been reduced. Moreover, the detailed calculation procedure and existence conditions of the desired observer have been presented as well. Finally, an illustrative example has been provided to show the effectiveness of the proposed algorithm. Future research issues would include the extension of our results to more complicated systems such as multi-rate systems, distributed systems, asynchronous sensor systems, and strict-feedback systems.

11
Conclusion and Further Work

The focus of the book has been placed on state estimation and fault diagnosis under imperfect measurements. Firstly, the phenomena under imperfect measurements have been introduced and the up-to-date results on analysis and synthesis of the systems subject to various factors have been reviewed. Then, in the following four chapters, the state estimation design methodologies for systems under imperfect measurements have been studied. Subsequently, the fault diagnosis problem has been taken into account.

Generally speaking, this book has established a unified theoretical framework for studying analysis and design problems under imperfect measurements while addressing difficulties induced by different factors such as nonlinear functions, missing measurements, transmission delays, signal quantizations, etc. It, however, should be stressed that the established results are still quite limited. Some of the related topics for the future research work are listed below:

- The majority of results about network-based FD/filtering problems usually make an explicit assumption that transmission rules (such as event-triggered transmission strategy and the quantization function) are determined in advance, based on which the filter/diagnoser is designed to ensure certain performance of the systems. It will be interesting to seek a novel method to co-design both the filter/diagnoser and the transmission rules in order to further improve the system performance.

- Though many network-induced phenomena have been introduced, most of the results are in fact obtained based on a centralized manner, i.e., the filter/diagnoser can access the whole measurement and be implemented accordingly. However, in a networked system, both the plant and the measurement may be spatially distributed, so another research frontier is to consider the estimation and diagnosis problems in a truly distributed setting.

- In order to make the addressed more safe and reliable, fault diagnosis is not enough in practical systems, some necessary sequential actions are needed after determining the location and amplitude of the fault. That is the basic task of fault-tolerant control (FTC). How to effectively combine the existing FD results and the FTC issue constitutes an interesting topic as well.

Bibliography

[1] S. Bakar, N. S. Ahmad, and P. Goh. Improved structured filter design and analysis for perturbed phase-locked loops via sector and H_∞ norm constraints with convex computations. *Computers and Electrical Engineering*, 81:106542, 2020.

[2] R. G. Agnel, and E. I. Jury. Almost sure boundedness of randomly sampled systems. *SIAM Journal on Control*, 9(3):372–384, 1971.

[3] H. Alikhani, and N. Meskin. Event-triggered robust fault diagnosis and control of linear Roesser systems: A unified framework? *Automatica*, 128:109575, 2021.

[4] M. D. S. Aliyu, and E. Boukas. Mixed H_2/H_∞ nonlinear filtering. *International Journal of Robust and Nonlinear Control*, 19(4):394–417, 2009.

[5] M. Basin, S. Elvira-Ceja, and E. Sanchez. Mean-square H_∞ filtering for stochastic systems: Application to a 2DOF helicopter. *Signal Processing*, 92(3):801–806, 2012.

[6] M. Basin, and P. Rodriguez-Ramirez. Sliding mode filter design for nonlinear polynomial systems with unmeasured states. *Information Sciences*, 204:82–91, 2012.

[7] J. Bellantoni, and K. W. Dodge. A square root formulation of the Kalman-Schmidt filter. *AIAA Journal*, 5(7):1309–1314, 1967.

[8] R. Bellman. *Introduction to Matrix Analysis*. New York: McGraw-Hill, 1970.

[9] S. Berkane, and A. Tayebi. Attitude estimation with intermittent measurements. *Automatica*, 105:415–421, 2019.

[10] K. Brammer. Input-adaptive Kalman-Bucy filtering. *IEEE Transactions on Automatic Control*, 15(1):157–158, 1970.

[11] R. Caballero-Águila, A. Hermoso-Carazo, J.D. Jiménez-López, J. Linares-Pérez, and S. Nakamori. Signal estimation with multiple delayed sensors using covariance information. *Digital Signal Processing*, 20(2):528–540, 2010.

[12] R. Caballero-Águila, A. Hermoso-Carazo, and J. Linares-Pérez. Optimal state estimation for networked systems with random parameter matrices, correlated noises and delayed measurements. *International Journal of General Systems*, 44(2):142–154, 2015.

[13] G. Calafiore. Reliable localization using set-valued nonlinear filters. *IEEE Transactions on Systems, Man and Cybernetics, Part A: Systems and Humans*, 35(2):189–197, 2005.

[14] G. C. Calafiore, and F. Abrate. Distributed linear estimation over sensor networks. *International Journal of Control*, 82(5):868–882, 2009.

[15] R. Carli, F. Fagnani, P. Frasca, and S. Zampieri. Gossip consensus algorithms via quantized communication. *Automatica*, 46(1):70–80, 2010.

[16] F. Carravetta, A. Germani, and M. Raimondi. Polynomial filtering for linear discrete time non-Gaussian systems. *SIAM Journal on Control and Optimization*, 34(5):1666–1690, 1996.

[17] M. G. Cea, and G. C. Goodwin. Event based sampling in non-linear filtering. *Control Engineering Practice*, 20(10):963–971, 2012.

[18] S. Challagundla, S. Chitraganti, and P. K. Wali. An efficient event-based state estimator for linear discrete-time system with multiplicative measurement noise. *IEEE Control Systems Letters*, 5(4):1315–1320, 2021.

[19] X.-H. Chang, Z.-M. Li, and J. H. Park. Fuzzy generalized H_2 filtering for nonlinear discrete-time systems with measurement quantization. *IEEE Transactions on Systems Man Cybernetics-Systems*. 48(12):2419-2430, 2018.

[20] X.-H. Chang, and Y. Liu. Robust H_∞ filtering for vehicle sideslip angle with quantization and data dropouts. *IEEE Transactions on Vehicular Technology*. 69(10):10435-10445, 2020.

[21] S. Chebotarev, D. Efimov, T. Raïssi, and A. Zolghadri. Interval observers for continuous-time LPV systems with L_1/L_2 performance. *Automatica*, 58:82–89, 2015.

[22] C.-K. Chen, and J.-H. Lee. Design of sharp-cutoff FIR digital filters with prescribed constant group delay. *IEEE Transactions on Circuits and Systems II-Analog and Digital Signal Processing*, 43(1):1–13, 1996.

[23] J. Chen, R. J. Patton, and H. Y. Zhang. Design of unknown input observers and robust fault-detection filters. *International Journal of Control*, 63(1):85–105, 1996.

[24] Y. Chen, Z. Wang, J. Hu, and Q.-L. Han. Synchronization control for discrete-time-delayed dynamical networks with switching topology under actuator saturations. *IEEE Transactions on Neural Networks and Learning Systems*, 32(5):2040–2053, 2021.

[25] Z. Cheng, H. Ren, B. Zhang, and R. Lu. Distributed Kalman filter for large-scale power systems with state inequality constraints. *IEEE Transactions on Industrial Electronics*, 68(7):6238–6247, 2021.

[26] G. Cong, F. Han, J. Li, and D. Dai. Event-triggered distributed filtering for discrete-time systems with integral measurements. *Systems Science and Control Engineering*, 9(1):272–282, 2021.

[27] H. Cox. On the estimation of state variables and parameters for noisy dynamic systems. *IEEE Transactions on Automatic Control*, 13(6):702–705, 1968.

[28] Q.K. Dang, and Y.S. Suh. Sensor saturation compensated smoothing algorithm for inertial sensor based motion tracking, *Sensors*, 14(5):8167–8188, 2014.

[29] J. Deyst, and C. Price. Conditions for asymptotic stability of the discrete minimum-variance linear estimator. *IEEE Transactions on Automatic Control*, 9(1):5–12, 1964.

[30] B. Ding. Stabilization of linear systems over networks with bounded packet loss and its use in model predictive control. *Automatica*, 47(11):2526–2533, 2011.

[31] D. Ding, Z. Wang, H. Dong, and H. Shu. Distributed H_∞ state estimation with stochastic parameters and nonlinearities through sensor networks: The finite-horizon case. *Automatica*, 48(8):1575–1585, 2012.

[32] D. Ding, Z. Wang, D. W. C. Ho, and G. Wei. Distributed recursive filtering for stochastic systems under uniform quantizations and deception attacks through sensor networks. *Automatica*, 78:231–240, 2017.

[33] H. Dong, Z. Wang, S. X. Ding, and H. Gao. Event-Based H_∞ filter design for a class of nonlinear time-varying systems with fading channels and multiplicative noises. *IEEE Transactions on Signal Processing*, 63(13):3387–3395, 2015.

[34] H. Dong, Z. Wang, and H. Gao. Robust H_∞ filtering for a class of nonlinear networked systems with multiple stochastic communication delays and packet dropouts. *IEEE Transactions on Signal Processing*, 58(4):1957–1966, 2010.

[35] H. Dong, Z. Wang, D. W. C. Ho, and H. Gao. Robust H_∞ filtering for Markovian jump systems with randomly occurring nonlinearities and sensor saturation: The finite-horizon case. *IEEE Transactions on Signal Processing*, 59(7):3048-3057, 2011.

[36] H. Dong, Z. Wang, and H. Gao. Distributed H_∞ filtering for a class of Markovian jump nonlinear time-delay systems over lossy sensor networks. *IEEE Transactions on Industrial Electronics*, 60(10):4665–4672, 2013.

[37] R. Duan, and J. Li. Finite-time distributed H_∞ filtering for Takagi-Sugeno fuzzy system with uncertain probability sensor saturation under switching network topology: Non-PDC approach. *Applied Mathematics and Computation*, 371:124961, 2020.

[38] D. Efimov, W. Perruquetti, T. Raïssi, and A. Zolghadri. Interval observers for time-varying discrete-time systems. *IEEE Transactions on Automatic Control*, 58(12):3218–3224, 2013.

[39] H. El Aiss, T. Zoulagh, A. El Hajjaji, and A. Hmamed. Full and reduced-order H_∞ filtering of Takagi-Sugeno fuzzy time-varying delay systems: Input-output approach. *International Journal of Adaptive Control and Signal Processing*, 35(5):748–768, 2021.

[40] S. Fan, H. Yan, H. Zhang, H. Shen, and K. Shi. Dynamic event-based non-fragile dissipative state estimation for quantized complex networks with fading measurements and its application. *IEEE Transactions on Circuits and Systems I-Regular Papers*, 68(2):856–867, 2021.

[41] P. Frasca, R. Carli, F. Fagnani, and S. Zampieri. Average consensus on networks with quantized communication, *International Journal of Robust and Nonlinear Control*, 19(16):1787–1816, 2009.

[42] P. J. Galkowski, and M. A. Islam. An alternative derivation of the modified gain function of Song and Speyer. *IEEE Transactions on Automatic Control*, 36(11):1323–1326, 1991.

[43] H. Gao, J. Lam, and C. Wang. Induced l_2 and generalized H_2 filtering for systems with repeated scalar nonlinearities. *IEEE Transactions on Signal Processing*, 53(11):4215–4226, 2005.

[44] Z. Gao, X. Liu, and M.Z.Q. Chen. Unknown input observer-based robust fault estimation for systems corrupted by partially decoupled disturbances. *IEEE Transactions on Industrial Electronics*, 63(4): 2537–3547, 2016.

[45] M. J. García-Ligero, A. Hermoso-Carazo, and J. Linares-Pérez. Least-squares estimators for systems with stochastic sensor gain degradation, correlated measurement noises and delays in transmission modelled by Markov chains. *International Journal of Systems Science*, 51(4):731–745, 2020.

[46] M. J. García-Ligero, A. Hermoso-Carazo, and J. Linares-Pérez. Distributed fusion estimation with sensor gain degradation and Markovian delays. *Mathematics*, 8(11):1948, 2020.

[47] W. Ge, and C.-Z. Fang. Detection of faulty components via robust observation. *International Journal of Control*, 47(2): 581–599, 1988.

[48] H. Geng, Z. Wang, L. Zou, A. Mousavi, and Y. Cheng. Protocol-based Tobit Kalman filter under integral measurements and probabilistic sensor failures. *IEEE Transactions on Signal Processing*, 69:546–559, 2021.

[49] E. Gershon, U. Shaked, and I. Yaesh. H_∞ *Control and Estimation of State-Multiplicative Linear Systems*. London, UK: Springer-Verlag London Limited, 2005.

[50] E. Gershon, U. Shaked, and N. Berman. H_∞ control and estimation of retarded state-multiplicative stochastic systems. *IEEE Transactions on Automatic Control*, 52(9):1773–1779, 2007.

[51] A. Germani, C. Manes, and P. Palumbo. Polynomial extended Kalman filtering for discrete-time nonlinear stochastic systems. In *Proceedings of the 42nd IEEE Conference on Decision and Control*, Maui, Hawaii USA, December, pages 886–891, 2003.

[52] A. Germani, C. Manes, and P. Palumbo. Polynomial extended Kalman filter. *IEEE Transactions on Automatic Control*, 50(12):2059–2064, 2005.

[53] A. Girard. Dynamic triggering mechanisms for event-triggered control. *IEEE Transactions on Automatic Control*, 60(7):1992–1997, 2015.

[54] Y. Guo, B. Huang. State estimation incorporating infrequent, delayed and integral measurements. *Automatica*, 58:32–38, 2015.

[55] J. Guo, Z. Wang, L. Zou, and Z. Zhao. Ultimately bounded filtering for time-delayed nonlinear stochastic systems with uniform quantizations under random access protocol. *Sensors*, 20(15):4134, 2020.

[56] W. Hahn. *Stability of Motion*. Berlin: Springer, 1967.

[57] S. Han, M. Xie, H. Chen, and Y. Ling. Intrusion detection in cyber-physical systems: Techniques and challenges. *IEEE Systems Journal*, 8(4):1049–1059, 2014.

[58] X. He, Z. Wang, Y. Liu, and D. H. Zhou. Least-squares fault detection and diagnosis for networked sensing systems using a direct state estimation approach. *IEEE Transactions on Industrial Informatics*, 9(3): 1670–1679, 2013.

[59] X. He, Z. Wang, and D. H. Zhou. Robust H_∞ filtering for time-delay systems with probabilistic sensor faults. *IEEE Signal Processing Letters*, 16(5):442–445, 2009.

[60] X. He, Z. Wang, and D. Zhou. Robust H_∞ filtering for networked systems with multiple state delays. *International Journal of Control*, 80(8):1217–1232, 2007.

[61] X. He, Z. Wang, and D. Zhou. State estimation for time-delay systems with probabilistic sensor gain reductions. *Asia-Pacific Journal of Chemical Engineering*, 3(6):712–716, 2008.

[62] X. He, Z. Wang, Y. Liu, and D. H. Zhou. Least-squares fault detection and diagnosis for networked sensing systems using a direct state estimation approach. *IEEE Transactions on Industrial Informatics*, 9(3):1670–1679, 2013.

[63] J. Hu, G.-P. Liu, H. Zhang, and H. Liu. On state estimation for nonlinear dynamical networks with random sensor delays and coupling strength under event-based communication mechanism. *Information Sciences*, 511:265–283, 2020.

[64] J. Hu, Z. Wang, F. E. Alsaadi, and T. Hayat. Event-based filtering for time-varying nonlinear systems subject to multiple missing measurements with uncertain missing probabilities. *Information Sciences*, 38:74–83, 2017.

[65] J. Hu, Z. Wang, and H. Gao. Joint state and fault estimation for time-varying nonlinear systems with randomly occurring faults and sensor saturations. *Automatica*, 97:150–160, 2018.

[66] J. Hu, Z. Wang, H. Gao, and L. K. Stergioulas. Extended Kalman filtering with stochastic nonlinearities and multiple missing measurements. *Automatica*, 48(9):2007–2015, 2012.

[67] S. Hu, and D. Yue. Event-based H_∞ filtering for networked system with communication delay. *Signal Processing*, 92(9):2029–2039.

[68] S. Hu, D. Yue, X. Yin, X. Xie, and Y. Ma. Adaptive event-triggered control for nonlinear discrete-time systems. *International Journal of Robust and Nonlinear Control*, 26(18):4104–4125, 2016.

[69] C. Huang, B. Shen, L. Zou, and Y. Shen. Event-triggering state and fault estimation for a class of nonlinear systems subject to sensor saturations. *Sensor*, 21(4):1242, 2021.

[70] J. Huang, Y. Wang, and T. Fukuda. Set-membership-based fault detection and isolation for robotic assembly of electrical connectors. *IEEE Transactions on Automation Science and Engineering*, 15(1):160–171, 2018.

[71] H. Ishii, and T. Basar. Remote control of LTI systems over networks with state quantization. *Systems and Control Letters*, 54(1):15–31, 2005.

[72] H. Jafari, and J. Poshtan. Event-triggered robust fault diagnosis Kalman filter for stochastic systems. *IEEE Sensors Journal*, 21(9):11031–11039, 2021.

[73] M. R. James, and I. R. Peterson. Nonlinear state estimation for uncertain systems with an integral constraint. *IEEE Transactions on Signal Processing*, 46(11):2926–2937, 1998.

[74] Y. Ji, W. Wu, H. Fu, and H. Qiao. Passivity-based filtering for networked semi-Markov robotic manipulators with mode-dependent quantization and event-triggered communication. *International Journal of Advanced Robotic Systems*, 18(2):1729881420939864, 2021.

[75] S. Julier, J. Uhlmann, and H. F. Durrant-Whyte. A new method for the nonlinear transformation of means and covariances in filters and estimators. *IEEE Transactions on Automatic Control*, 45(3):477–482, 2000.

[76] R. E. Kalman. Nonlinear aspects of sampled-data control systems. *Proceedings of the Symposium on Nonlinear Circuit Analysis*, 6:273–313, 1956.

[77] R. E. Kalman. A new approach to linear filtering and prediction problems. *Transactions of the ASME Journal of Basic Engineering*, 82(1):35–45, 1960.

[78] S. Kar, B. Sinopoli, and J. Moura. Kalman filtering with intermittent observations: Weak convergence to a stationary distribution, *IEEE Transactions on Automatic Control.* 57(2):405–420, 2012.

[79] H. K. Khalil. *Nonlinear Systems* (Third Edition). New Jersey: Prentice Hall, 2002

[80] U. A. Khan, and J. M. F. Moura. Distributing the Kalman filter for large-scale systems. *IEEE Transactions on Signal Processing*, 56(10):4919–4935, 2008.

[81] R. Z. Khasminskii. *Stochastic Stability of Differential Equations*. The Netherlands: Sijtjoff and Noordhoff, Alphen aan den Rijn, 1980.

[82] R. A. Kisner, W. W. Manges, L. P. MacIntyre, J. J. Nutaro, J. K. Munro, P. D. Ewing, M. Howlader, P. T. Kuruganti, R. M. Wallace, and M. M. Olama. *Cybersecurity through Real-Time Distributed Control Systems*. Oak Ridge, TN, USA: Oak Ridge National Laboratory, 2014.

[83] S. Kluge, K. Reif, and M. Brokate. Stochastic stability of the extended Kalman filter with intermittent observations. *IEEE Transactions on Automatic Control*, 55(2):514–518, 2010.

[84] D. Koenig, N. Bedjaoui, and X. Litrico. Unknown input observers design for time-delay systems application to an open-channel. *Proceedings of 44th European Control Conference on Decision and Control, Seville, Spain, December 2005*, pages 5794–5799.

[85] K. Kowalski and W. H. Steeb. *Nonlinear Dynamical Systems and Carleman Linearization*. Singapore: World Scientific, 1991.

[86] H. Li, Z. Chen, L. Wu, H.-K. Lam, and H. Du. Event-triggered fault detection of nonlinear networked systems. *IEEE Transactions on Cybernetics*, 47(4):1041–1052, 2017.

[87] X. Li, and H. Gao. Robust finite frequency H_∞ filtering for uncertain 2-D Roesser systems. *Automatica*, 48(6):1163–1170.

[88] Z. Li, J. Hu, and J. Li. Distributed filtering for delayed nonlinear system with random sensor saturation: A dynamic event-triggered approach. *Systems Science and Control Engineering*, 9(1):440–454, 2021.

[89] P. Li, J. Lam, and G. Chesi. On the synthesis of linear H_∞ filters for polynomial systems. *System and Control Letters*, 61(1):31–36, 2012.

[90] S. Li, Z. Li, J. Li, T. Fernando, H. H.-C. Iu, Q. Wang, and X. Liu. Application of event-triggered cubature Kalman filter for remote nonlinear state estimation in wireless sensor network. *IEEE Transactions on Industrial Electronics*, 68(6):5133–5145, 2021.

[91] N. Li, Q. Li, and J. Suo. Dynamic event-triggered H_∞ state estimation for delayed complex networks with randomly occurring nonlinearities. *Neurocomputing*, 421:97–104, 2021.

[92] Y. Li, S. Liu, Y. Li, and D. Zhao. Fault estimation for discrete time-variant systems subject to actuator and sensor saturations. *International Journal of Robust and Nonlinear Control*, 31(3):988–1004, 2021.

[93] P. Li, and Y. Shen. Anti-disturbance adaptive sampled-data observers for a class of nonlinear systems with unknown hysteresis. *International Journal of Robust and Nonlinear Control*, 31(8):3212–3229, 2021.

[94] H. Li, and Y. Shi. Robust H_∞ filtering for nonlinear stochastic systems with uncertainties and Markov delays. *Automatica*, 48(1):159–166, 2012.

[95] J. Li, Z. Wang, H. Dong, and G. Ghinea. Outlier-resistant remote state estimation for recurrent neural networks with mixed time-delays. *IEEE Transactions on Neural Networks and Learning Systems*, 32(5):2266–2273, 2021.

[96] D. Liang, Y. Yang, R. Li, and R. Liu. Finite-frequency H_-/H_∞ unknown input observer-based distributed fault detection for multi-agent systems. *Journal of the Franklin Institute-Engineering and Applied Mathematics*, 358(6):3258–3275, 2021.

[97] W. Liu, G. Tao, and C. Shen. Robust measurement fusion steady-state estimator design for multisensor networked systems with random two-step transmission delays and missing measurements. *Mathematics and Computers in Simulation*, 181:242–283, 2021.

[98] Q. Liu, Z. Wang, X. He, G. Ghinea, and F. E. Alsaadi. A resilient approach to distributed filter design for time-varying systems under stochastic nonlinearities and sensor degradation. *IEEE Transactions on Signal Processing*, 65(5):1300–1309, 2017.

[99] Q. Liu, Z. Wang, X. He, and D. H. Zhou. Event-based recursive distributed filtering over wireless sensor networks. *IEEE Transactions on Automatic Control*, 60(9):2470–2475, 2015.

[100] Q. Liu, Z. Wang, Q.-L. Han, and C. Jiang. Quadratic estimation for discrete time-varying non-Gaussian systems with multiplicative noises and quantization effects. *Automatica*, 113:108714, 2020.

[101] S. Liu, Z. Wang, J. Hu, and G. Wei. Protocol-based extended Kalman filtering with quantization effects: The Round-Robin case. *International Journal of Robust and Nonlinear Control*, 30(18):7927–7946, 2020.

[102] D. Liu, Z. Wang, Y. Liu, and F. E. Alsaadi. Recursive filtering for stochastic parameter systems with measurement quantizations and packet disorders. *Applied Mathematics and Computation*, 398:125960, 2021.

[103] S. Liu, Z. Wang, G. Wei, and M. Li. Distributed set-membership filtering for multirate systems under the Round-Robin scheduling over sensor networks. *IEEE Transactions on Cybernetics*, 50(5):1910–1920, 2020.

[104] J. Liu, M. Yang, X. Xie, C. Peng, and H. Yan. Finite-time H_∞ filtering for state-dependent uncertain systems with event-triggered mechanism and multiple attacks. *IEEE Transactions on Circuits and Systems I-Regular Papers*, 67(3):1021–1034, 2020.

[105] J. Liu, T. Yin, M. Shen, X. Xie, and J. Cao. State estimation for cyber-physical systems with limited communication resources, sensor saturation and denial-of-service attacks. *ISA Transactions*, 104:101–114, 2020.

[106] A. Liu, W.-A. Zhang, B. Chen, and L. Yu. Networked filtering with Markov transmission delays and packet disordering. *IET Control Theory and Applications*, 12(5):687–693, 2018.

[107] E. B. Lopez-Montero, J. Wan, and O. Marjanovic. Trajectory tracking of batch product quality using intermittent measurements and moving window estimation. *Journal of Process Control*, 25:115–128, 2015.

[108] J. Lu, W. Wang, L. Li, and Y. Guo. Unscented Kalman filtering for nonlinear systems with sensor saturation and randomly occurring false data injection attacks. *Asian Journal of Control*, 23(2):871–881, 2021.

[109] D. G. Luenberger. An introduction to observers. *IEEE Transactions on Automatic Control*, 16(6):596–602, 1971.

[110] L. Ma, Z. Wang, H.-K. Lam, and N. Kyriakoulis. Distributed event-based set-membership filtering for a class of nonlinear systems with sensor saturations over sensor networks. *IEEE Transactions on Cybernetics*, 47(11):3772–3783, 2017.

[111] S. Makni, M. Bouattour, A. El Hajjaji, and M. Chaabane. Robust fault estimation and fault-tolerant tracking control for uncertain Takagi-Sugeno fuzzy systems: Application to single link manipulator. *International Journal of Adaptive Control and Signal Processing*, 35(5):846–876, 2021.

[112] A. Manor, A. Osovizky, E. Dolev, E. Marcus, D. Ginzburg, V. Pushkarsky, Y. Kadmon, and Y. Cohen. Compensation of scintillation sensor gain variation during temperature transient conditions using signal processing techniques. In *IEEE Nuclear Science Symposium Conference*, Orlando, FL, USA, October, pages 2399–2403, 2009.

[113] F. Mazenc, and O. Bernard. Asymptotically stable interval observers for planar systems with complex poles. *IEEE Transactions on Automatic Control*, 55(2):523–527, 2010.

[114] F. Mazenc, and O. Bernard. Interval observers for linear time-invariant systems with disturbances. *Automatica*, 47(1):140–147, 2011.

[115] X. Meng, and T. Chen. Event triggered robust filter design for discrete-time systems. *IET Control Theory and Applications*, 8(2): 104–113, 2014.

[116] M. Miskowicz. Asymptotic effectiveness of the event-based sampling according to the integral criterion. *Sensors*, 7(1):16–37, 2007.

[117] M. Miskowicz. Send-on-delta concept: An event-based data reporting strategy. *Sensors*, 6(1):49–63, 2006.

[118] H. Modares, F. L. Lewis, and M. B. Naghibi-Sistani. Adaptive optimal control of unknown constrained-input systems using policy iteration and neural networks. *IEEE Transactions on Neural Networks and Learning Systems*, 24(10):1513–1525, 2013.

[119] E. Mousavinejad, X. Ge, Q.-L. Han, T. J. Lim, and L. Vlacic. An ellipsoidal set-membership approach to distributed joint state and sensor fault estimation of autonomous ground vehicles. *IEEE-CAA Journal of Automatica Sinica*, 8(6):1107–1118, 2021.

[120] Z. Nagy, Z. Lendek, and L. Busoniu. Observer design for a class of nonlinear systems with nonscalar-input nonlinear consequents. *IEEE Control Systems Letters*, 5(3):971–976, 2021.

[121] V. H. Nguyen, and Y. S. Suh. Improving estimation performance in networked control systems applying the send-on-delta transmission method. *Sensors*, 7(10):2128–2138, 2007.

[122] V. H. Nguyen, and Y. S. Suh. Networked estimation with an area-triggered transmission method. *Sensors*, 8(2):897–909, 2008.

[123] C. M. Nguyen, C. P. Tan, and H. Trinh. Sliding mode observer for estimating states and faults of linear time-delay systems with outputs subject to delays. *Automatica*, 124:109274, 2021.

[124] V. Nithya, R. Sakthivel, and Y. Ren. Resilient H_∞ filtering for networked nonlinear Markovian jump systems with randomly occurring distributed delay and sensor saturation. *Nonlinear Analysis-Modelling and Control*, 26(2):187–206, 2021.

[125] D. Nodland, H. Zargarzadeh, and S. Jagannathan. Neural network based optimal adaptive output feedback control of a helicopter UAV. *IEEE Transactions on Neural Networks and Learning Systems*, 24(7):1061–1073, 2013.

[126] R. Olfati-Saber. Distributed Kalman filtering for sensor networks. *Proceedings of the 46th IEEE Conference on Decision and Control*, New Orleans, LA, USA, December, pages 1814–1820, 2007.

[127] M. Pourasghar, V. Puig, and C. Ocampo-Martinez. Interval observer versus set-membership approaches for fault detection in uncertain systems using zonotopes. *International Journal of Robust and Nonlinear Control*, 29(10):2819–2843, 2019.

[128] J. Qi, and Y. Li. Event-triggered L_1 filtering for uncertain networked control systems with multiple sensor fault modes. *Transactions of the Institute of Measurement and Control*, 43(6):1325–1336, 2021.

[129] N. Qian, and G. Chang. Optimal filtering for state space model with time-integral measurements. *Measurement*, 176:109209, 2021.

[130] W. Qian, Y. Li, G. Chen, and W. Liu. L_2-L_∞ filtering for stochastic delayed systems with randomly occurring nonlinearities and sensor saturation. *International Journal of Systems Science*, 51(13):2360–2377, 2020.

[131] J. Qiu, H. Gao, and S.X. Ding. Recent advances on fuzzy-model-based nonlinear networked control systems: A survey. *IEEE Transactions on Industrial Electronics*, 63(2):1207–1217, 2016.

[132] B. Qu, Z. Wang, and B. Shen. Fusion estimation for a class of multi-rate power systems with randomly occurring SCADA measurement delays. *Automatica*, 125:109408, 2021.

[133] F. Rahimi, and H. Rezaei. An event-triggered recursive state estimation approach for time-varying nonlinear complex networks with quantization effects. *Neurocomputing*, 426:104–113, 2021.

[134] T. Raïssi, D. Efimov, and A. Zolghadri. Interval state estimation for a class of nonlinear systems. *IEEE Transactions on Automatic Control*, 57(1):260–265, 2012.

[135] B. Rao, and H. Durrant-Whyte. Fully decentralized algorithm for multisensor Kalman filtering. *IEE Proceedings-D: Control Theory and Applications*, 138:413–420, 1991.

[136] K. Reif, S. Gunther, E. Yaz, and R. Unbehauen. Stochastic stability of the discrete-time extended Kalman filter. *IEEE Transactions on Automatic Control*, 44(4):714–728, 1999.

[137] H. Ren, R. Lu, J. Xiong, and Y. Xu. Optimal estimation for discrete-time linear system with communication constraints and measurement quantization. *IEEE Transactions on Systems Man Cybernetics-Systems*, 50(5):1932–1942, 2020.

[138] H. Rezaei, R. M. Esfanjani, A. Akbari, and M. H. Sedaaghi. Scalable event-triggered distributed extended Kalman filter for nonlinear systems subject to randomly delayed and lost measurements. *Digital Signal Processing*, 111:102957, 2021.

[139] F. C. Schweppe. Recursive state estimation: Unknown but bounded errors and system inputs. *IEEE Transactions on Automatic Control*, 13(1):22–28, 1968.

[140] U. Shaked, and V. Suplin. A new bounded real lemma representation for the continuous-time case. *IEEE Transactions on Automatic Control*, 46(9):1420–1426, 2001.

[141] C. Shang, S. X. Ding, and H. Ye. Distributionally robust fault detection design and assessment for dynamical systems. *Automatica*, 125:109434, 2021.

[142] B. Shen, Z. Wang, H. Shu, and G. Wei. Robust H_∞ finite-horizon filtering with randomly occurred nonlinearities and quantization effects. *Automatica*, 46(11):1743–1751, 2010.

[143] B. Shen, Z. Wang, and Y. S. Hung. Distributed H_∞-consensus filtering in sensor networks with multiple missing measurements: The finite-horizon case. *Automatica*, 46(10):1682–1688, 2010.

[144] Y. Shen, Z. Wang, B. Shen, and F. E. Alsaadi. H_∞ filtering for multi-rate multi-sensor systems with randomly occurring sensor saturations under thep-persistent CSMA protocol. *IET Control Theory and Applications*, 14(10):1255–1265, 2020.

[145] D. Shi, T. Chen, and L. Shi. An event-triggered approach to state estimation with multiple point- and set-valued measurements. *Automatica*, 50(6):1641–1648, 2014.

[146] J. Sijs, and M. Lazar. Event based state estimation with time synchronous updates. *IEEE Transactions on Automatic Control*, 57(10):2650–2655, 2012.

[147] B. Sinopoli, L. Schenato, M. Franceschetti, K. Poolla, M. Jordan, and S. Sastry. Kalman filtering with intermittent observations. *IEEE Transactions on Automatic Control*, 49(9):1453–1464, 2004.

[148] I. S. D. Solomon, and A. J. Knight. Spatial processing of signals received by platform mounted sonar. *IEEE Journal of Oceanic Engineering*, 27(1):57–65, 2002.

[149] T. L. Song, and J. L. Speyer. A stochastic analysis of a modified gain extended Kalman filter with application to estimation with bearing only measurement. *IEEE Transactions on Automatic Control*, 30(10):940–949, 1985.

[150] W. Song, Z. Wang, J. Wang, F. Alsaadi, and J. Shan. Particle filtering for nonlinear/non-Gaussian systems with energy harvesting sensors subject to randomly occurring sensor saturations. *IEEE Transactions on Signal Processing*, 69:15–27, 2021.

[151] X. Song, M. Wang, B. Zhang, and S. Song. Event-triggered reliable H_∞ fuzzy filtering for nonlinear parabolic PDE systems with Markovian jumping sensor faults. *Information Sciences*, 510:50–69, 2020.

[152] Q. Su, Z. Fan, T. Lu, Y. Long, and J. Li. Fault detection for switched systems with all modes unstable based on interval observer. *Information Sciences*, 517:167–182, 2020.

[153] Y. S. Suh, V. H. Nguyen, and Y. S. Ro. Modified Kalman filter for networked monitoring systems employing a send-on-delta method. *Automatica*, 43(2):332–338, 2007.

[154] W. Sun, M. Qiu, and X. Lv. H_∞ filter design for a class of delayed Hamiltonian systems with fading channel and sensor saturation. *AIMS Mathematics*, 5(4):2909–2922, 2020.

[155] Y. Sun, J. Mao, H. Liu, and D. Ding. Distributed event-based set-membership filtering for a class of nonlinear systems with sensor saturations over sensor networks. *Neurocomputing*, 400:412–419, 2020.

[156] S. Sun, L. Xie, W. Xiao, and N. Xiao. Optimal filtering for systems with multiple packet dropouts. *IEEE Transactions on Circuits and Systems*, 55(7):695–699, 2008.

[157] H. Tan, B. Shen, Y. Liu, A. Alsaedi, and B. Ahmad. Event-triggered multi-rate fusion estimation for uncertain system with stochastic nonlinearities and colored measurement noises. *Information Fusion*, 36:313–320, 2017.

[158] T. J. Tarn, and Y. Rasis. Observers for nonlinear stochastic systems. *IEEE Transactions on Automatic Control*, 21(4):441–448, 1976.

[159] H.-C. Ting, J.-L. Chang, and Y.-P. Chen. Proportional-derivative unknown input observer design using descriptor system approach for non-minimum phase systems. *International Journal of Control, Automation, and Systems*, 9(5):850–856, 2011.

[160] J. M. Velni, and K. M. Grigoriadis. Delay-dependent H_∞ filtering for time-delayed LPV systems. *Systems and Control Letters*, 57(4):290–299, 2008.

[161] Z. Wang, H. Dong, B. Shen, and H. Gao. Finite-Horizon H_∞ filtering with missing measurements and quantization effects. *IEEE Transactions on Automatic Control*, 58(7):1707–1718, 2013.

[162] T. Wang, H. Gao, and J. Qiu. A combined adaptive neural network and nonlinear model predictive control for multirate networked industrial process control. *IEEE Transactions on Neural Networks and Learning Systems*, 27(2):416–425, 2016.

[163] Z. Wang, D. Hong, B. Shen, and H. Gao. Finite-horizon H_∞ filtering with missing measurements and quantization effects. *IEEE Transactions on Automatic Control*, 58(7):1707-1718, 2013.

[164] Z. Wang, B. Shen, and X. Liu. H_∞ filtering with randomly occurring sensor saturations and missing measurements. *Automatica*, 48(3):556–562.

[165] J. Wang, Y. Shi, M. Zhou, Y. Wang, and V. Puig. Active fault detection based on set-membership approach for uncertain discrete-time systems. *International Journal of Robust and Nonlinear Control*, 30(14):5322–5340, 2020.

[166] S. Wang, Z. Wang, H. Dong, and Y. Chen. Distributed state estimation under random parameters and dynamic quantizations over sensor networks: A dynamic event-based approach. *IEEE Transactions on Signal and Information Processing over Networks*, 6: 732–743, 2020.

[167] F. Wang, Z. Wang, J. Liang, and X. Liu. Resilient filtering for linear time-varying repetitive processes under uniform quantizations and Round-Robin protocols. *IEEE Transactions on Circuits and Systems I-Regular Papers*, 65(9):2992–3004, 2018.

[168] Z. Wang, F. Yang, D. W. C. Ho, and X. Liu. H_∞ filtering for stochastic time-delay systems with missing measurements. *IEEE Transactions on Signal Process*, 54(7):2579–2587, 2006.

[169] Y. Wang, W. X. Zheng, and H. Zhang. Dynamic event-based control of nonlinear stochastic systems. *IEEE Transactions on Automatic Control*, 62(12):6544–6551, 2017.

[170] H. Witsenhausen. Sets of possible states of linear systems given perturbed observations. *IEEE Transactions on Automatic Control*, 13(5):556–558, 1968.

[171] G. Wei, Z. Wang, and B. Shen. Error-constrained filtering for a class of nonlinear time-varying delay systems with non-gaussian noises. *IEEE Transactions on Automatic Control*, 55(12):2876–2882, 2010.

[172] J. Wu, Q.-S. Jia, K. H. Johansson, and L. Shi. Event-based sensor data scheduling: Trade-off between communication rate and estimation quality. *IEEE Transactions on Automatic Control*, 58(4):1041–1046, 2013.

[173] L. Wu, and D. W. C. Ho. Reduced-order L_2/L_∞ filtering of switched nonlinear stochastic systems. *IET Control Theory and Applications*, 3(5):493–508, 2009.

[174] J. Wu, G. Shi, B. D. O. Anderson, and K. H. Johansson. Kalman filtering over Gilbert-Elliott channels: Stability conditions and critical curve. *IEEE Transactions on Automatic Control*, 63(4):1003–1017, 2018.

[175] L. Wu, and Z. Wang. Fuzzy filtering of nonlinear fuzzy stochastic systems with time-varying delay. *Signal Processing*, 89(9):1739–1753, 2009.

[176] L. Xie, Y. C. Soh, and C. E. de Souza. Robust Kalman filtering for uncertain discrete-time systems. *IEEE Transactions on Automatic Control*, 39(6):1310–1314, 1994.

[177] J. Xiong, and J. Lam. Stabilization of linear systems over networks with bounded packet loss. *Automatica*, 43(1):80–87, 2007.

[178] K. Xiong, C. Wei, and L. Liu. Robust extended Kalman filtering for nonlinear systems with stochastic uncertainties. *IEEE Transactions on Systems, Man, and Cybernetics-Part A: Systems and Humans*, 40(2):399–405.

[179] K. Xiong, L. Liu, and Y. Liu. Robust extended Kalman filtering for nonlinear systems with multiplicative noises. *Optimal Control Applications and Methods*, 32(1):47–63, 2011.

[180] J. Xu, L. Sheng, and M. Gao. Fault estimation for nonlinear systems with sensor gain degradation and stochastic protocol based on strong tracking filtering. *Systems Science and Control Engineering*, 9:60–70, 2021.

[181] S. Xu, J. Lam, and X. Mao. Delay-dependent H_∞ control and filtering for uncertain Markovian jump systems with time-varying delays. *IEEE Transactions on Circuits and Systems I-Regular Papers*, 54(9):2070–2077, 2007.

[182] Y. Xu, R. Lu, P. Shi, J. Tao, and S. Xie. Robust estimation for neural networks with randomly occurring distributed delays and Markovian jump coupling. *IEEE Transactions on Neural Networks and Learning Systems*, 29(4):845–855, 2018.

[183] H. Yalcin, R. Collins, and M. Hebert. Background estimation under rapid gain change in thermal imagery. *Computer Vision and Image Understanding*, 106(2-3):148–161, 2007.

[184] H. Yang, H. Shu, Z. Wang, F. E. Alsaadi, and T. Hayat. Almost sure state estimation with H_2-type performance constraints for nonlinear hybrid stochastic systems. *Nonlinear Analysis-Hybrid Systems*, 19:26–37, 2016.

[185] R. Yang, P. Shi, and G.-P. Liu. Filtering for discrete-time networked nonlinear systems with mixed random delays and packet dropouts. *IEEE Transactions on Automatic Control*, 56(11):2655–2660, 2011.

[186] L. Yi, Y. Liu, W. Yu, and J. Zhao. A novel nonlinear observer for fault diagnosis of induction motor. *Journal of Algorithms and Computational Technology*, 14:1748302620922723, 2020.

[187] H. Yu, Y. Zhuang, and W. Wang. Distributed H_∞ filtering with consensus in sensor networks: A two-dimensional system-based approach. *International Journal of Systems Science*, 42(9):1543–1557, 2011.

[188] C. Zammali, J. Van Gorp, Z. Wang, and T. Raïssi. Sensor fault detection for switched systems using interval observer with L_∞ performance. *European Journal of Control*, 57:147–156, 2021.

[189] X. Zhang. Sensor bias fault detection and isolation in a class of nonlinear uncertain systems using adaptive estimation. *IEEE Transactions on Automatic Control*, 56(5):1220–1226, 2011.

[190] J. Zhang, P. D. Christofides, X. He, Z. Wu, Z. Zhang, and D. Zhou. Event-triggered filtering and intermittent fault detection for time-varying systems with stochastic parameter uncertainty and sensor saturation. *International Journal of Robust and Nonlinear Control*, 28(16):4666–4680, 2018.

[191] J. Zhang, S. Gao, G. Li, J. Xia, X. Qi, and B. Gao. Distributed recursive filtering for multi-sensor networked systems with multi-step sensor delays, missing measurements and correlated noise. *Signal Processing*, 181:107868, 2021.

[192] L. Zhang, H. Liang, Y. Sun, and C. K. Ahn. Adaptive event-triggered fault detection scheme for semi-Markovian jump systems with output quantization. *IEEE Transactions on Systems Man Cybernetics-Systems*, 51(4):2370–2381, 2021.

[193] X. Zhang, M. Polycarpou, and T. Parisini. Robust fault isolation for a class of non-linear input-output systems. *International Journal of Control*, 74(13):1295–1310, 2001.

[194] Y. Zhang, Z. Wang, and F. E. Alsaadi. Detection of intermittent faults for nonuniformly sampled multi-rate systems with dynamic quantisation and missing measurements. *International Journal of Control*, 93(4):898–909, 2020.

[195] Y. Zhang, Z. Wang, L. Zou, and Z. Liu. Fault detection filter design for networked multi-rate systems with fading measurements and randomly occurring faults. *IET Control Theory and Applications*, 10(5):573–581, 2016.

[196] Z.-H. Zhang and G.-H. Yang. Distributed fault detection and isolation for multiagent systems: An interval observer approach. *IEEE Transactions on Systems, Man, and Cybernetics: Systems*, 50(6):2220–2230, 2020.

[197] Z. Zhao, Z. Wang, L. Zou, and J. Guo. Set-membership filtering for time-varying complex networks with uniform quantisations over randomly delayed redundant channels. *International Journal of Systems Science*, 51(16):3364–3377, 2020.

[198] Y. Zheng, H. Fang, and H. O. Wang. Takagi-Sugeno fuzzy-model-based fault detection for networked control systems with Markov delays. *IEEE Transactions on Systems, Man, and Cybernetics Part B-Cybernetics*, 36(4):924–929, 2006.

[199] M. Zhong, T. Xue, X. Zhu, and L. Zhang. An H_i/H_∞ optimisation approach to distributed event-triggered fault detection over wireless sensor networks. *International Journal of Systems Science*, 52(6):1160–1170, 2021.

[200] D. H. Zhou, X. He, Z. Wang, G.-P. Liu, and Y. D. Ji. Leakage fault diagnosis for an internet-based three-tank system: An experimental study. *IEEE Transactions on Control Systems Technology*, 20(4):857–870, 2012.

[201] B. Zhou, K. Qian, X. Ma, and X. Dai. Ellipsoidal bounding set-membership identification approach for robust fault diagnosis with application to mobile robots. *Journal of Systems Engineering and Electronics*, 28(5):986–995, 2017.

[202] D. Zhou, L. Qin, X. He, R. Yan, and R. Deng. Distributed sensor fault diagnosis for a formation system with unknown constant time delays. *Science China-Information Sciences*, 61(11):112205, 2018.

[203] M. Zhou, Z. Cao, M. Zhou, and J. Wang. Finite-frequency H_-/H_∞ fault detection for discrete-time T-S fuzzy systems with unmeasurable premise variables. *IEEE Transactions on Cybernetics*, 51(6):3017–3026, 2021.

[204] X. Zhou, and Z. Gu. Event-triggered H_∞ filter design of T-S fuzzy systems subject to hybrid attacks and sensor saturation. *IEEE Access*, 8:126530–126539, 2020.

[205] X. Zhu, and D. Li. Robust fault estimation for a 3-DOF helicopter considering actuator saturation. *Mechanical Systems and Signal Processing*, 155:107624, 2021.

[206] X. Zhu, Y. Liu, J. Fang, and M. Zhong. Fault detection for a class of linear systems with integral measurements. *Science China-Information Sciences*, 64(3):132207, 2021.

[207] L. Zou, Z. Wang, J. Hu, and D. Zhou. Moving horizon estimation with unknown inputs under dynamic quantization effects. *IEEE Transactions on Automatic Control*, 65(12):5368–5375, 2020.

Index

Adaptive threshold, 90–92, 108, 111, 113
Asymptotic stability in probability, 23, 62, 67–70, 72–74, 78

Bernoulli distributed white sequence, 8, 9, 21, 63, 80, 117

Closed-loop system, 8, 18, 151, 153, 156, 158, 160
Control input, 19, 25, 75, 80, 82, 91, 136, 145, 149–152, 155, 165
Cost function, 20, 141, 144

Data dropout, 3, 27, 29, 115
Distributed filtering, 14, 16, 21, 22, 39
Disturbance attenuation level, 9, 14, 19, 64, 66, 68, 70–76

Euclidean norm, 46, 108, 111, 113, 144, 160
Event-triggered transmissions (ETT), 17, 24, 25, 115–117, 123, 133, 134, 138, 141, 145
Exponential boundedness, 5, 20
Extended Kalman filter (EKF), 4, 6, 8, 12, 13, 89, 90, 96

Fault diagnosis (FD), xv, 23, 90, 149, 150, 155, 158, 159, 162, 163, 177
Fault estimation, xvi, 5, 8, 15, 19, 118, 163, 164, 175
Filter gain, 4, 12, 17, 20, 21, 24, 40, 91, 96, 98, 99, 117, 120, 127, 128, 136, 140

Filtering error, xvi, 7, 11–14, 16–21, 24, 33, 36, 40, 44, 97, 113, 116, 119

Gaussian distributions, 89, 115, 133
Gaussian white noise, 4, 56, 62, 89

H_∞ filtering, 8, 13, 19, 115, 133, 137
Imperfect measurements, 23, 177
Integral measurement, 3, 22, 164–166

Kalman filter, 4, 36, 37, 61, 115, 133

Linear matrix inequality (LMI), 1, 5, 18, 134, 137, 164
Linearization error, 1, 6, 89
Lipschitz condition, 1, 6, 135
Local measurement, 10, 43
Logarithmic quantizer, 11, 25, 135, 145
Luenberger observer, 4, 149, 156, 162

Markov chain, 9, 18, 62, 63, 76, 78
Mathematical induction, 30, 41
Measurement noise, 16, 28, 108, 134, 150, 165
Minimum variance, 1, 23, 39, 45, 59, 116
Missing measurements, xv, 24, 46, 79, 84, 177
Moore-Penrose pseudo inverse, 46, 143
Multi-agent systems, 6, 79, 163
Multiplicative noise, 10, 12, 16, 41

Index

Nonlinear function, 1, 7, 11, 24, 63, 68, 91–93, 96, 113, 116, 135, 177
Nonlinear system, xvi, 4, 6–9, 12, 15, 19–21, 61, 65, 72, 74, 82, 89–91, 111, 113, 116, 123, 132, 134, 137, 144, 162–164

Optimal filter, 4, 36, 37, 141

Polynomial approximation, xvi, 6, 89, 92, 96, 103
Polynomial extended Kalman filter (PEKF), 6, 89, 90, 94, 111
Posterior density function (PDF), 115, 119, 123

Random variable, 12, 23, 28, 41, 59
Residual evaluation, 90, 159
Residual signal, 1, 4, 5, 22, 90, 96, 103, 151
Riccati difference equation, 13, 25, 133, 141

Sector-bounded condition, 1, 7, 19
Sensor failures, 3, 21, 61, 79, 108, 166
Sensor gain degradation, xvi, 9, 14, 23, 27, 36, 40, 51, 55

Sensor network, xvi, 16, 23, 39, 44, 46, 56
Sensor saturation, 19, 61, 76, 129
Signal quantization, 2, 11, 17
Signal transmissions, 3, 8, 115, 148, 155
Signum function, 19, 63, 117
Spectral norm, 51, 108
State estimation, xv, 22, 25, 43, 79, 89, 113, 128, 159, 164, 166, 171, 175
Stochastic process, xvi, 46, 47, 50, 123, 127
Stochastic stability, 9, 13, 17, 20
Stochastic systems, 1, 8, 9, 11–13, 17, 20, 61

Time delays, 7, 23, 28, 37, 61, 68, 70, 74, 78, 159
Time-invariant systems, 5, 39, 137
Time-varying systems, 16, 18, 22, 23, 27, 40, 59, 84, 103, 134, 148, 159

Unknown input observer (UIO), 6, 149, 160, 175
Upper bound, 11, 16, 18, 20, 91, 96, 97, 100, 101, 105, 113, 118, 128, 132